BBNJ

国际协定草案
要素概览

（二）

郑苗壮 等◎编译

中国社会科学出版社

图书在版编目(CIP)数据

BBNJ 国际协定草案要素概览. 二／郑苗壮等编译. —北京：中国社会科学出版社，2020.4

ISBN 978-7-5203-6099-9

Ⅰ.①B… Ⅱ.①郑… Ⅲ.①国际海域—海洋生物资源—生物多样性—生物资源保护—国际条约—文件—汇编 Ⅳ.①P745②D993.5

中国版本图书馆 CIP 数据核字(2020)第 036980 号

出 版 人	赵剑英	
责任编辑	梁剑琴	
责任校对	沈丁晨	
责任印制	郝美娜	

出　　版	中国社会科学出版社	
社　　址	北京鼓楼西大街甲 158 号	
邮　　编	100720	
网　　址	http://www.csspw.cn	
发 行 部	010-84083685	
门 市 部	010-84029450	
经　　销	新华书店及其他书店	

印刷装订	北京市十月印刷有限公司	
版　　次	2020 年 4 月第 1 版	
印　　次	2020 年 4 月第 1 次印刷	

开　　本	880×1230 1/32	
印　　张	6.375	
插　　页	2	
字　　数	148 千字	
定　　价	48.00 元	

序

国家管辖范围以外区域海洋生物多样性（BBNJ）养护和可持续利用问题国际文书谈判政府间大会于 2018 年 9 月正式开启，标志着有关 BBNJ 的国际立法进程进入新阶段。各国正朝着在《联合国海洋法公约》框架下制定第三个执行协定的方向努力迈进。

BBNJ 国际文书是当前海洋法领域最重要的国际立法进程，涉及海洋遗传资源的获取和惠益分享、海洋保护区等划区管理工具、环境影响评价、能力建设和海洋技术转让等重要问题，攸关海洋新资源和海洋活动空间等战略利益，牵动国际海洋秩序的调整和变革，各国对国际文书谈判予以高度重视。

学界和实务界关于 BBNJ 问题的讨论由来已久，2004 年该问题被正式纳入联合国议程，同年第 59 届联大设立 BBNJ 问题特设工作组，讨论和研究加强 BBNJ 养护和可持续利用的可行方案。历经 11 年九次特设工作组会议的讨论磋商，联大于 2015 年 6 月 19 日通过 69/292 号决议，决定在《联合国海洋法公约》框架下就 BBNJ 问题制订具有法律约束力的国际文书，并建立谈判预备委员会，就国际文书草案的要素进行讨论并向联大提出实质建议。预备委员会历时两年，先后举行四次会议，于 2017 年 7 月通过了向联大提交了要素建议草案，同时声明该草案不代表各方就相关要素达成的共识，其不影响各国在未来政府间大会谈判中的立场。在此基础上，联大于 2017 年 12 月 24 日通过第 72/249 号决议，决定在联合国主持下于 2018 年至 2020 年上半年首先召开四次政府间大会，就制订有关国际文书展开政府间谈判。

中国作为快速发展中的海洋大国，高度重视国家管辖范围以外区域的海洋生物多样性国际立法问题，积极参与了历次特设工作组会议、谈判预备委员会会议以及政府间大会首次会议。我本人作为中国代表团团长以及预备委员会和政府间大会首次会议的主席团成员，参加了预委会所有四次会议和政府间大会首次会议。会上，中国代表团深度参与了 BBNJ 有关问题的讨论和磋商，积极发出中国声音，努力贡献中国智慧和方案，在不少问题上发挥引领作用，特别是在 2018 年 9 月政府间大会首次会议上的发言全面而富有建设性，有效扩大了中国在有关进程中的话语权和影响力。中国代表团还两度以中国政府名义就国际文书草案要素提交书面意见，系统阐述中国对有关问题的立场主张和法律依据，对中国参与和引导会议进程发挥了重要作用。

在前一阶段参与相关磋商和谈判过程中，包括自然资源部海洋发展战略研究所同仁在内的一批国内专家学者组成强有力的研究团队，对 BBNJ 国际文书所涉法律问题开展深入研究，积极建言献策，为中国政府代表团参与有关谈判提供了有力支撑。为了帮助学界和实务界及时跟踪了解相关立法进程，海洋发展战略研究所的同事们对预委会主席精简文件进行了编译，旨在反映 BBNJ 国际文书谈判进程的阶段性情况，以推动相关问题的深入研究。

　　相信本书对于国内学界和实务界进一步研究 BBNJ 问题以及中国政府代表团参与后续政府间谈判具有重要的参考价值。在此，谨对有关同事的努力和付出表示衷心感谢！

马新民①

---

　　① 作者系时任外交部条法司副司长、BBNJ 预委会和政府间大会第一次会议中国代表团团长，现任中国驻苏丹大使。

目　　录

Chair's streamlined non-paper on elements of a draft text of an international legally-binding instrument under the United Nations Convention on the Law of the Sea on the conservation and sustainable use of marine biological diversity of areas beyond

国家管辖范围以外区域海洋生物多样性国际协定草案要素概览（二）

联大第 69/292 号决议所设立的预备委员会第三次会议指出：要根据《联合国海洋法公约》的规定制定一份具有法律约束力的国际文书，以实现国家管辖范围以外区域海洋生物多样性的养护与可持续利用，文件是 2017 年 2 月 28 日发布的概览（一）的精简版本。本国际文书是基于各国提交的主场文件以及会议讨论的观点及其补充文件编制的，并考虑了各代表团提出的提案、问题和意见。

概览（二）可作为一个参考文件来协助各代表团审议预备委员会处理的各项问题。它对各代表团提出的意见和建议进行了汇编，但未考虑对这些意见和建议提供何种程度上的支持。在本文件中列入这些意见和建议并不意味着各代表团已就这些意见和建议达成一致或统一意见。如有多个选择方案，其在本文件中的出现顺序不应被解释为这些方案存在先后次序。

概览（二）的内容不影响任何代表团对所提及的任何事项的立场。此外，概览（二）所列出的要素并不详尽，因此不排除对其中未包含的事项加以考虑。

主席在此感谢对本非纸质文件的编写提出建议、提案和意见的代表团。①

---

① 相关内容，参见 http：//www. un. org/depts/los/biodiversity/prepcom_ files/ rolling_ comp/Submissions_ StreamlinedNP. pdf。

一 序言要素

1. 对背景问题的描述，如：

• 海洋生物多样性对海洋健康、生产力和恢复力、粮食安全、生态系统服务以及当代和后代的可持续发展的重要性。

• 对养护和可持续利用国家管辖范围以外区域海洋生物多样性的重要性。

• 气候变化与海洋之间的联系。

• 环境影响评价在预防和识别对海洋环境造成潜在威胁的作用。

• 使用划区管理工具（如海洋保护区）来有效保护和养护海洋生物多样性的重要性。

• 能力建设和技术转让对各国，特别是发展中国家的重要性。

2. 对行动原因的说明，如：

• 需建立一个综合的全球机制以更好地处理国家管辖范围以外区域海洋生物多样性养护和可持续利用问题。

• 需与现有相关机构进行密切合作和协调。

• 希望就国家管辖范围以外区域海洋生物多样性养护和可持续利用情况建立一个有效机制，包括建立一个公平公正的海洋遗传资源获取和惠益分享机制。

● 海洋遗传资源获取和惠益分享机制的法律确定性非常重要。

3. 承认《联合国海洋法公约》为所有海洋活动提供了法律框架。

4. 参考并承认根据《联合国鱼类种群协定》等相关国际文书，以及国际海事组织、国际海底管理局、区域渔业管理组织和区域性海洋管理组织等相关组织所开展的工作。

5. 重申沿海国对其大陆架，包括超过 200 海里以外大陆架的管辖权和主权权利。

6. 也可在新国际文书序言中列出一些方法和原则，如：

● 本国际文书不应损害现有相关法律文书和框架以及相关全球、区域和部门机构。

● 人类共同继承财产。

● 公海自由。

● 特别考虑发展中国家，尤其是小岛屿发展中国家和最不发达国家。

● 全球和区域合作的重要性。

● 各国公平和公正地分享利用海洋遗传资源产生的惠益。

● 人类共同关切事项。

二　总则要素

A. 术语的使用①

7. 定义需与《联合国海洋法公约》《联合国鱼类种群协定》《生物多样性公约》及其《关于获取遗传资源和公正公平分享其利用所产生惠益的名古屋议定书》（《名古屋议定书》）和其他相关国际文书中的定义一致，并应根据国家管辖范围以外区域海洋生物多样性的具体情况进行调整。

8. 定义并不涵盖商品贸易。

9. 术语和定义可包括：

● 国家管辖范围以外区域

Ø "国家管辖范围以外区域"是指在《联合国海洋法公约》中所界定的公海和国际海底区域。

● 划区管理工具

Ø 方案1：划区管理工具的定义可包括三个要素：（1）目标——"划区管理工具"的目标是实现海洋生物多样性养护和可持续利用；（2）地理范围——"划区管理工具"只适用于公海和国际海底区域；（3）功能——划区管理工具可包括不同的

―――――――――

① 可在概览（二）的相应章节给出术语定义，除非这些术语在概览（二）的多个章节出现。

功能和管理方法。

Ø 方案 2："划区管理工具"是指为实现既定目标（环境养护或/和资源管理）指定和应用位于国家管辖范围以外特定区域的工具。

Ø 方案 3：是指为实现特定目标，而对某一地理区域进行空间管理的工具。通过该工具可实现对以一个或多个部门/活动的管理，并提供比周边区域更高级别的保护。

Ø 方案 4：划区管理工具包括促进海洋生物多样性养护和可持续利用的部门和跨部门措施。跨部门划区管理工具，包括海洋保护区和海洋空间规划，需要多个组织和机构进行合作与协调，从而实现更广泛的目标并对累积影响作出响应。部门划区管理工具包括主管国际组织为特定区域实现生物多样性养护目标而采取的措施，包括区域渔业管理组织指定的深海脆弱生态系统、国际海事组织指定的特别敏感海域、国际海底管理局指定的特别环境利益区（参照区）。

- 生物多样性

Ø "生物多样性"是指各种来源的活生物体的变异性，这些来源除其他外，包括陆地、海洋和其他水生生态系统及其所构成的生态综合体；这包括物种内部、物种之间和生态系统的多样性。

- 生物资源

Ø "生物资源"是指对人类具有实际或潜在用途或价值的遗传资源、生物体或其部分、生物种群或生态系统中任何其他生物组成部分。

- 生物勘探
- 生物技术

Ø "生物技术"是指使用海洋生物系统、活生物体或其衍生物的技术应用，目的是制造或改进产品或加工用于特殊使用。

• 大陆架，见《联合国海洋法公约》中的定义。[①]

• 衍生物

Ø "衍生物"是指由生物资源或遗传资源的遗传表达或新陈代谢产生的天然生物化石化合物，即使其不具备遗传功能单位（根据《名古屋议定书》第2条）。

• 生态系统

Ø "生态系统"指植物、动物和微生物群落及其非生物环境作为功能单位相互作用形成的一个动态复合体。

• 基于生态系统的管理

Ø "基于生态系统的管理"是指通过一体化的方法管理整个生态系统，包括考虑所有利益攸关者及其活动，以及由此引起的对所考虑的生态系统产生直接或间接影响的压力及其来源。基于生态系统的管理旨在通过制定和实施跨部门生态系统层面的管理计划，维持或重建生态系统健康，以及富有生产力和弹力的环境。

• 环境影响评价

Ø "环境影响评价"是指通过考虑对相关社会经济、文化和人类健康造成的有利和不利影响，对在国家管辖范围以外区域进行的、对国家管辖范围以内或以外区域有影响的活动所造成的环境影响进行评估的过程。

• 非原生境获取

---

① 沿海国的大陆架包括其领海以外依其陆地领土的全部自然延伸，扩展到大陆边外缘的海底区域的海床和底土，如果从测算领海宽度的基线量起到大陆外缘的距离不到二百海里，则扩展到二百海里的距离。

- 遗传材料

Ø 方案 1："遗传材料"是指来自植物、动物、微生物或其他来源的任何含有遗传功能单位的材料（根据《生物多样性公约》第 2 条）。

Ø 方案 2："遗传材料"是指来自植物的包含遗传功能单位的材料，包括生殖和营养繁殖物质（根据《粮食和农业植物遗传资源国际条约》第 2 条）。

Ø 方案 3："遗传材料"是指从"区域"采集的包含遗传功能单位的植物、动物或微生物来源的材料；它不包括由衍生物等制成的材料或说明材料的信息，如遗传序列的变化信息。

- （海洋）遗传资源

Ø 定义须考虑用作遗传性质研究的鱼类和用作商品的鱼类之间的区别。对于其他动物物种（如可用作商品的软体动物），也存在相同区别。

Ø 定义可包括以下要素：（1）海洋中动物、植物、微生物或其他来源；（2）包含遗传功能单位的遗传材料；（3）共存实际或潜在价值；（4）来自国家管辖范围以外区域。

Ø 方案 1："遗传资源"是指具有实际或潜在价值的遗传材料。

Ø 方案 2："海洋遗传资源"是指从"区域"内的植物、动物或微生物中采集的具有实际或潜在价值的海洋遗传材料。

Ø 方案 3："海洋遗传资源"是指来自植物、动物、微生物或其他来源，包含遗传功能单位，并具有实际或潜在价值的海洋遗传材料。

- 生物信息的获取
- 原生境获取

Ø "原地采集"是指在国家管辖范围以外区域的生态系统和自然栖息地采集海洋遗传材料。

· 海洋保护区

Ø 该定义可把海洋保护区划为划区管理工具（旨在实现对海洋生物多样性和生态系统的长期养护）的一个子类别。

Ø 定义需足够宽泛或灵活，以便把区域渔业管理组织已建立的公海保护区包括在内，也就是说，可根据新国际文书将其充分认定为海洋保护区。

Ø 方案1："海洋保护区"是指为实现特定养护目标而指定、控制和管理的特定海洋区域（《生物多样性公约》第2条）。

Ø 方案2：《生物多样性公约》第2条中关于"保护区"的定义仅仅是一个起始点，需特别针对国家管辖范围以外的海洋区域进行调整。

Ø 方案3："海洋保护区"是指为实现特定养护目标（包括进行长期养护和提高自然恢复力）而在国家管辖范围以外区域指定的、对人类活动进行管制和管理或禁止人类活动的海洋区域。

Ø 方案4："海洋保护区"是指所界定的海洋环境区域，包括与其相关的植物群落、动物群落和已通过立法或其他有效手段（包括传统知识）保留的历史和文化特征，且对其海洋生物多样性的保护高于周围区域。

Ø 方案5：指"为实现生物多样性养护或渔业管理目的，比周边水域享受更高级别保护的任何海洋区域"。海洋保护区不限于海洋保留区或禁捕区。

· 海洋科学研究

· 海洋空间规划

Ø 海洋空间规划是一个跨部门划区管理工具，《联合国海洋法公约》为海洋有序和可持续利用提供了框架，以平衡发展和保护海洋环境需求。部门划区管理工具（如禁渔区、特别敏感海域和特别环境利益区）、其他跨部门划区管理工具（如海洋保护区）、战略环境影响评价和环境影响评价，是这一总体规划方法的组成部分。海洋空间规划是基于生态系统的具有适应性的方法，且包括所考虑的区域内的所有利益攸关方。

- 可持续利用

Ø "可持续利用"是指以不会导致生物多样性长期下降的方式和速度使用海洋生物多样性的组成部分，以保持其潜力满足当代和后代需求和愿望。

- 技术

Ø "技术"是指硬技术及所有相关方面，如专门设备和专业技术知识，包括组装、维护和操作可行系统所必需的手册、设计、操作说明、培训和技术咨询与协助，以及为此目的非排他性地使用以上信息的合法权利。它也指使此类转让成为可持续转让的基础设施和提升技术能力。

- 跨界环境影响评价
- 海洋技术转让

Ø 海洋技术转让是指为增进对海洋及其资源的研究和理解，对获取和运用知识所需的仪器、设备、船舶、流程和方法进行的转让。

- 利用海洋遗传资源

Ø 方案 1："利用海洋遗传资源"是指对海洋遗传资源的遗传成分和/或生物化学成分进行研究和开发，包括通过运用《生物多样性公约》第 2 条所界定的生物技术。

Ø 方案 2："海洋遗传资源利用"是指对遗传资源的遗传成分和/或生物化学成分进行商业研究和开发，包括通过运用生物技术。

B. 范围/应用

1. 地理范围

10. 方案 1：国家管辖范围以外区域。

11. 方案 2：现有国际公约未充分规定的区域。

12. 不适用于国家管辖范围内的海域，包括超过 200 海里以外的大陆架。

2. 属事范围

13. 联合国大会第 69/292 决议中"一揽子方案"的所有要素。

14. 关于活动：

• 方案 1：养护、可持续利用和负责任管理国家管辖范围以外区域所有海洋生物。

• 方案 2：在国家管辖范围以外区域进行的受某一缔约方管辖或控制的活动。

• 方案 3：任何可能对国家管辖范围以外区域的海洋生物多样性，包括海洋进程产生影响的活动或开发。

• 方案 4：有可能对国家管辖范围以外区域的海洋生物多样性或生态系统造成重大影响或破坏的活动（不论这些活动发生在何处）。

• 方案 5：所有在国家管辖范围以外区域进行的，且/或可能对国家管辖范围以外区域的海洋生物多样性和资源产生影响的活动。如果此类活动受现有国际文书的管理或管辖，则新国际文书将采用现有国际文书的相关规定，并作出必要变更。

● 方案 6：就"一揽子方案"中所确定的要素，对国家管辖范围以外区域海洋生物多样性产生影响，而不损害现有相关法律文书和框架以及相关全球、区域和部门机构的现有和新兴活动。

● 方案 7：现有国际公约，如《联合国海洋法公约》和《生物多样性公约》等未充分规定的活动。

● 方案 8：国家管辖范围以外的渔业管理不构成谈判的一部分。

3. 属人范围

15. 新国际文书将与《联合国鱼类种群协定》一样，把参与范围扩大至所有国家和其他实体。

C. 目标

16. 确保实现国家管辖范围以外区域海洋生物多样性养护和可持续利用。

17. 其他潜在目标包括：

● 保护和保全海洋环境。

● 加强区域合作和区域合作机制。

● 防止或消除产能过剩，并确保所涉实体的努力水平与生物多样性可持续利用之间的相称性，从而确保养护和可持续管理措施的有效性。

D. 与《联合国海洋法公约》及其他文书和框架以及相关全球、区域和部门机构的关系

18. 与《联合国海洋法公约》的关系：

● 新国际文书中的所有规定皆不应损害《联合国海洋法公约》规定各国的权利、管辖权和义务。新国际文书的释义和适用应与《联合国海洋法公约》相一致。

19. 与其他文书的关系：

• 新国际文书不应损害现有相关法律文书和框架以及相关全球性、区域性和部门机构。

Ø 方案 1："互不抵触"条款将有助于实现这一目标，其中包括与《联合国鱼类种群协定》第 44 条类似的规定，即不影响其他条约约定的权利和义务。

Ø 方案 2：新国际文书不会影响相关国际组织和协会在其职权范围内的权限。

• 明确说明新国际文书在国家管辖范围以外区域的活动（如根据《东北大西洋海洋环境保护公约》进行的活动）方面不具备的作用或功能。

Ø 方案 3：根据《联合国海洋法公约》第 208 条第 3 款的规定，新国际文书和管理机构所实施的条例和措施的效力"不得低于国际规则、标准和建议的办法及程序"。

Ø 方案 4：国家管辖范围以外区域采用的标准不得低于专属经济区的标准。

20. 新国际文书的规定不适用于享有主权豁免权的船舶（根据《联合国海洋法公约》第 236 条的规定）。

21. 《联合国海洋法公约》或新国际文书未规范事项将继续以一般国际法规则和原则执行。

三 国家管辖范围以外区域海洋生物多样性的养护和可持续利用

A. 一般原则和方法①

22. 区分原则和方法。

23. 指导原则和方法的定义和/或解释需与《联合国海洋法公约》《联合国鱼类种群协定》《生物多样性公约》以及1992年《里约环境与发展宣言》等其他相关国际文书中商定的一致。

24. 原则和方法部分的内容可包括：

- 认识到需建立一个全面的全球制度，以更好地处理国家管辖范围以外区域海洋生物多样性的养护和可持续利用。

- 尊重《联合国海洋法公约》所载的权利、义务利益和自由的平衡。

- 纳入且不损害《联合国海洋法公约》所载的有关原则。

- 人类共同继承财产。

- 公海自由。

- 认可现有相关法律文书和框架以及相关全球、区域和部

---

① 引用方法和原则可包括：1）在文书中明确引用这些方法和原则［在概览（二）序言部分或在一篇独立文章中引用；某些方法和原则可能适合在其本身所在的文章中进行进一步阐述，如《联合国鱼类种群协定》第6条和第7条］；2）通过实施，在概览（二）的各项规定中对这些原则和方法加以反映。

门机构（特别是《联合国海洋法公约》《联合国鱼类种群协定》、区域渔业管理组织/协定、国际海事组织、国际海底管理局和区域海洋环境保护公约）。

- 不损害现有相关法律文书和框架以及相关全球、区域和部门机构。
- 适当考虑他国的权利。
- 尊重沿海国对其国家管辖范围内的区域所拥有的所有权利，包括对超过 200 海里以外的大陆架（如适用）的权力。
- 尊重沿海国的主权和领土完整。
- 兼容性。
- 毗邻和咨询毗邻国要求。
- 认可毗邻沿海国家以及其他国家的作用。
- 加强国家和组织之间的合作与协调，从而实现对国家管辖范围以外区域海洋生物多样性养护和可持续利用。
- 保护和养护海洋环境及其生物多样性，为子孙后代造福。
- 不要将一种污染转化为另一种污染，也不要转移损害或危害。
- 只为和平目的利用国家管辖范围以外区域的生物多样性。
- 一体化方法。
- 生态系统方法。
- 基于科学的方法。
- 使用最佳可获科学信息。
- 信息公开。
- 公众参与。
- 利益攸关方参与。
- 善治。

- 透明度。
- 利用传统知识。
- 问责。
- 平等。
- 代内和代际公平。
- 能力建设和技术转让。
- 为防止或减轻对生物多样性的损害，所采用的环保技术和操作方法。
- 海洋生物多样性可持续利用。
- 预防原则/方法。
- "基于风险"的方法。
- 污染者付费原则。
- 小岛屿发展中国家和最不发达国家等发展中国家的特殊利益、情况和需求。
- 避免不成比例的负担。
- 适应性管理。
- 解决累积影响的能力。
- 可追溯性。
- 灵活性。
- 保护生物多样性是人类共同关注的问题。

25. 根据新国际文书，气候变化影响是作出决定及采取行动时需考虑的因素之一，且所作出的决定不应加剧或加速气候变化的不利影响，特别是对小岛屿发展中国家的影响。

26. 根据新国际文书采取的行动或活动不应被解释或认为有损于缔约国对任何领土或海洋主权争端或海洋区域划界争议所持的立场。

B. 国际合作①

27. 将通过信息交流等方式加强国家和国际组织，包括区域和部门机构之间的合作、协调、协商和沟通。

28. 新国际文书可以目标、程序、准则、标准和指南的形式，包括通过参与国家管辖范围以外海洋生物多样性养护和可持续利用的现有全球、部门或区域组织，向各国提供指导和建议。

29. 对出现的新问题（如气候变化、海洋酸化或污染对鱼类种群的不利影响）加以考虑，并经商定，提出海洋生物多样性保护的通用指南或方法。

30. 各国有义务在考虑次区域或区域具体特点的基础上，直接或通过适当的次区域、区域或全球机制进行合作（见《联合国鱼类种群协定》第8.1条）。

31. 应在执行新国际文书时，各缔约国应通力协作并积极推动主管国际组织在其职权范围内采取行动，从而为实现新国际文书的目标作出贡献。如果某一缔约国认为相关主管国际组织应就其职权范围内的某一问题采取行动，该缔约国需向有权解决该问题的组织提出该问题。该问题涉及的组织或安排中的相关缔约方，应与该组织或安排一道进行合作，以便作出适当响应。

32. 如果某一缔约国提出应由对海洋生物多样性有管理权的一个政府间组织来采取某项行动，当该行动可能对主管部门或区域组织已制定的保护和管理措施产生重大影响时，该缔约国应通过该组织与其成员或参与者协商。在可行的情况下，此类

---

① 另见第四节。

协商应在向政府间组织提交提案之前进行。

33. 可针对现有组织制定业绩激励方案，包括通过扩大现在组织的职权，明确授予它们在国家管辖范围以外区域采取海洋生物多样性养护措施的权力、为现有组织的治理框架增加新的原则、在现有组织内部制定执行新国际文书相关部分的程序，或制定谅解备忘录，来促进与其他组织的协调。

34. 如果没有任何部门或机构，被授权在特定区域或国家管辖范围以外区域对海洋生物多样性进行保护和养护：

• 方案 1：鼓励设立有关机构。

• 方案 2：不鼓励设立有关机构，因为这不在新国际文书的范围内。

35. 在存在若干机构但缺少有效协调机制的情况下，可鼓励在一定时限内建立有效的协调机制（《联合国鱼类种群协定》第8.5 条）。

36. 可加强或发展正式或非正式区域合作机制。

37. 预计有关主管国际组织在其职权范围内参与和/或就根据新国际文书作出的实际安排进行合作的可能性。

38. 国际组织的协调和合作义务可通过以下方式进一步落实：

• 适时参加相关会议；

• 就国家管辖范围以外区域的事项进行磋商，以期协调各自的活动；

• 在收集国家管辖范围以外区域的数据和信息方面进行合作；

• 与新国际文书设立的科学机构分享有关活动的信息和数据以及其职权范围内活动影响的相关数据；

● 就在特定时限内确定和执行最有效的保护措施进行合作；

● 合作管理国家管辖范围以外的海洋保护区；

● 进行海洋科学研究并对现有海洋保护区及其保护措施的有效性进行联合评估；

● 定期向缔约国大会报告进展情况；

● 作为观察员参加各管理机构的会议。

C. 海洋遗传资源包括惠益分享

1. 范围

39. 地理范围：

● 方案1：新国际文书将适用于"区域"和公海的海洋遗传资源。

● 方案2：新国际文书只适用于"区域"内的海洋遗传资源。

40. 实质范围：

● 用于研究遗传性质的鱼类和其他生物资源。

Ø 设定一个基于科学信息的阈值；如果为海洋遗传资源生物勘探的目的所获取或捕获的某类（鱼类）物种的数量超过一定量（取决于物种和栖息地变异性），则其将被视为商品。科学/技术机构可根据新国际文书对该阈值进行阐述。

● 关于原生境/非原生境/生物信息学方式获取的资源：

Ø 方案1：适用于原生境获取和非原生境获取海洋遗传资源。

Ø 方案2：适用于原生境获取和非原生境获取海洋遗传资源以及以生物信息学和数字序列数据获取的资源。

Ø 方案3：适用于原生境获取的海洋遗传资源。

● 衍生物：

Ø 方案 1：适用于衍生物。

Ø 方案 2：不适用于衍生物。

2. 指导原则和方法

41. 关于人类共同继承财产和公海自由：

● 方案 1：对人类共同继承财产的认识是制定国家管辖范围以外区域海洋遗传资源新管理制度的基石。这意味着：

Ø 所开展的活动，不论其地理位置在哪里，都应符合整个人类的利益，并应特别考虑发展中国家的利益和需求。

Ø 任何对国家管辖范围以外区域的主权主张或行使或划拨行为皆不被认可。

Ø 将以公平和公正的方式分享惠益。

Ø 所述区域内的有关资源勘探和开采活动将受到新国际文书的管理。

● 方案 2：公海自由原则将适用于国家管辖以外区域的海洋遗传资源。

● 方案 3：人类共同关切事项。

● 方案 4：未指明所适用的法律制度。

42. 各国应仅为和平目的利用国家管辖范围以外的区域及其资源。

43. 需尊重沿海国对其大陆架［包括超过 200 海里以外的大陆架（如适用）］所拥有的管辖权和权利。

44. 毗邻原则。沿海国家可在袋状区域的资源养护以及对其遗传资源获取的管理和规范方面发挥更大作用。

45. 根据《联合国海洋法公约》第 241 条的规定，海洋科学研究活动不应构成对海洋环境的任何部分或其资源进行权力主张的法律依据。

46. 传统知识。

47. 法律确定性、清晰度和透明度。

48. 鼓励研究、创新和商业开发。

49. 遗传材料的可持续采集。

50. 环保技术和操作方法。

51. 公平和非任意惠益分享规则和程序。

52. 简单、便利和具有成本效益的程序和机制。

3. 获取和惠益分享

53. 可考虑现有获取和惠益分享模式，包括：

- 《联合国海洋法公约》有关海洋科学研究的规定。

- 《联合国海洋法公约》第 82 条。

- 《生物多样性公约》和《名古屋议定书》。

- 《粮食和农业植物遗传资源国际条约》。

- 《南极条约体系》。

54. 新国际文书下的科学/技术机构可制定并向全球机构提出关于获取和惠益分享制度准则的建议。

3.1 国家管辖范围以外区域海洋遗传资源的获取

55. 关于是否对获取进行管制：

- 方案 1：根据《联合国海洋法公约》中关于国家管辖范围以外区域海洋科学研究的规定，自由获取海洋遗传资源。

- 方案 2：对获取进行管制：

Ø 方案 2.1：为生物勘探，而非海洋科学研究目的的获取。

Ø 方案 2.2：对"区域"内海洋遗传资源的获取。

Ø 可根据国际海底管理局的模式，设定获取条款和条件，同时考虑改变用途的可能性，包括能力建设、海洋技术转让、样本存放要求，可从开放平台，如数据库、生物数据库和/或生物样本库获取的数据和相关信息，作为获取遗传资源的条件，对获取和惠益分享基金的捐款。

Ø 可参考《名古屋议定书》中关于遗传资源以及涉及土著人民和当地社区事先同意的原则。

Ø 要求各国采取适当和有效的法律、行政或政策措施，来保证在其管辖范围内按照既定规则来利用遗传资源。

56. 可在以下条件下，原生境获取国家管辖范围以外区域的海洋遗传资源：

- 尊重《联合国海洋法公约》下的海洋科学研究制度。
- 不妨碍研究和开发。
- 按照《联合国海洋法公约》的规定，尊重沿海国对其管辖下的资源享有的权利和义务。
- 按照《联合国海洋法公约》相关规定，对国家管辖范围以外区域海洋生物多样性进行养护和可持续利用。
- 船旗国以不损害生态系统的方式，即采用环保技术和操作方法，来采集海洋遗传资源。
- 考虑对毗邻区域，包括国家管辖区域的影响。
- 考虑对毗邻缔约国的影响程度。
- 考虑补偿和防治污染措施的成本。
- 如需要，在海洋保护区中采取更严格的环境保护措施。
- 根据《粮食和农业植物遗传资源国际条约》第 16 条，制定发展中国家参与的关于国家管辖范围以外区域海洋环境养护和管理的海洋科学研究项目。

3.2 海洋遗传资源利用惠益分享

3.2.1 目标

57. 促进对国家管辖范围以外区域海洋生物多样性的养护和

可持续利用。

58. 有利于当代和后代。

59. 促进海洋科学研究。

60. 加强研究和开发。

61. 促进能力建设和技术转让。

62. 就获取国家管辖范围以外区域的海洋遗传资源进行能力建设。

3.2.2 惠益分享指导原则

63. 惠益分享指导原则可包括：

• 平衡参与国和其他海洋遗传资源获取和利用实体之间的利益。

• 基于人类共同遗产的认识，实现公平公正。

• 保持透明。

• 促进对国家管辖范围以外区域海洋生物多样性的养护和可持续利用。

• 对各国根据《联合国海洋法公约》的规定进行海洋科学研究的权利没有负面影响。

• 有利于根据《联合国海洋法公约》进行海洋科学研究，并可促进知识创造和技术创新，且不会对研究与开发产生损害。

• 适当考虑小岛屿发展中国家和最不发达国家。

• 增加在生物多样性保护方面的科学知识。

• 利用衍生物产生的惠益，且不会允许利用衍生物来破坏或损害人类生命或用于非和平目的。

3.2.3 惠益

64. 惠益类型可包括：

• 方案1：货币和非货币形式。

23

- 方案 2：非货币性惠益。

65. 货币性惠益可包括《名古屋议定书》附件以及《粮食和农业植物遗传资源国际条约》第四部分提到的惠益。

66. 非货币性惠益可包括：

- 《名古屋议定书》附件中提到的惠益。
- 《粮食和农业植物遗传资源国际条约》第四部分提到的惠益。
- 促进海洋科学研究。
- 在海洋科学研究以及研究与开发项目方面的合作。
- 通过数据分享机制（如信息交换所、数据库、样本库和开放性基因库等）获取和传播所有形式的资源、样本、数据和相关知识。
- 传播与海洋遗传资源有关的研究和开发成果。
- 采集和分享与海洋环境、生物多样性和生态系统相关的数据和知识。
- 新国际文书可就国家管辖范围以外区域的海洋遗传资源，规定一个指定、协调、促进和监测第 13 部分（"海洋科学研究"）所载条款执行的框架（如促进在海洋科学研究领域的国际合作（《联合国海洋法公约》第 242 条）、出版和传播从海洋科学研究获得的知识〔《联合国海洋法公约》第 244 条第 1 款）以及促进数据和信息流动和知识转让（《联合国海洋法公约》第 244 条第 2 款）〕。
- 技术转让，包括根据《联合国海洋法公约》第 14 部分进行技术转让。
- 能力建设，包括邀请发展中国家科学家参与科学研究、提供科学研究船舶、教育和培训项目，开展以加强、促进和鼓励材料、信息和知识分享为目的的活动、加强技术转让能力、进行机构能力建设以及人力资源和材料管理（增强管理和实施相关获取规定的能力）、设立全球奖学基金，以及建立区域中心。

- 其他社会经济效益（如针对健康和安全等优先需求的研究）。

67. 可考虑在进程中的某个时间点进行分享惠益的类型。

3.2.4 惠益分享方式

68. 可考虑建立把现有惠益分享机制（包括国际海底管理局）考虑在内的机制。

69. 可建立信息交换所机制。①

70. 可采用的惠益分享模式包括：

- 可以在可能的信息交换所机制内设立一项基金：

Ø 可在新国际文书中对该基金的用途进行一般性描述，管理机构或通过议定书制定该基金详细的运作模式。

Ø 通过特许权使用费或分阶段付款获得资金。

Ø 最不发达国家将是该基金的主要受益者。

Ø 可对小岛屿发展中国家提供具体拨款。

Ø 资金将用于国家管辖范围以外区域海洋生物多样性的养护和可持续利用。

Ø 可考虑是否给予有义务向惠益分享基金捐资的发展中国家和最不发达国家特别豁免权。

- 可建立与世界卫生组织《共享流感病毒以及获得疫苗和其他利益的大流行性流感防范框架》下的年度合作伙伴贡献类

---

① 请参阅本节第 5 小节。

似的制度。

● 可制定国际海洋研究计划，支持国家管辖范围以外区域海洋环境的养护和管理。根据《粮食和农业植物遗传资源国际条约》第 5 部分（支持成分）的条款，研究计划可包括关于发展中国家参与的规定。可把这些方案与现有国家研究机构联系起来，并就国家管辖范围以外区域的遗传材料开展研究活动，从而为参与者带来货币性惠益。也可要求发达国家邀请发展中国家参与海洋遗传资源研究和开发活动。

● 可通过把基于项目的方法（与《粮食和农业植物遗传资源国际条约》中的惠益分享基金类似）与从《共享流感病毒以及获得疫苗和其他利益的大流行性流感防范框架》获得的灵感结合起来，建立一种混合机制。

3.3 知识产权

71. 方案 1：知识产权包括专利申请中的强制来源披露要求，不应包括在新国际文书的讨论范围，而应由负责该议题的现有机构（世界知识产权组织和世界贸易组织）解决。

72. 方案 2：可提出在专利或其他知识产权申请时对海洋遗传资源来源进行强制性披露的要求。

73. 方案 3：新国际文书将禁止私自盗用和行使知识产权，但这将限制为进一步研究或其他目的而获得海洋遗传资源。如有任何一方就从海洋遗传资源开发的产品提出知识产权请求，可考虑采用《粮食和农业植物遗传资源国际条约》中的方法。

74. 方案 4：知识产权问题的解决方式必须与世界知识产权组织所采取的方式相一致。

75. 方案 5：制定一个特殊制度。

4. 对国家管辖范围以外区域海洋遗传资源利用情况的监测

76. 用户需注册所开展的活动。

77. 可通过制定议定书、行为规范或指南，来确保在国家管辖范围以外区域利用海洋遗传资源的透明度。

78. 海洋遗传资源开采信息库可用来追踪在国家管辖范围以外区域获得的海洋遗传资源的来源。

79. 可为国家管辖范围以外区域的海洋遗传资源引入"证书"。该证书可参考《名古屋议定书》下的"国际文书"制度，将伴随遗传资源，作为其在研究、开发、创新或商业化阶段的来源证明。

80. 可对国际海底管理局在监测海洋遗传资源利用方面的职能进行规定。

81. 可建立负责国家知识产权的国家主管部门，以对监测海洋遗传资源的利用情况进行监测，并确保惠益分享。

5. 信息交换所机制①

82. 可把《名古屋议定书》获取和惠益分享信息交换所作为模板，并进一步扩大其内容，以包括与国家管辖范围以外区域海洋生物多样性有关的记录和数据。

83. 信息交换所机制可：

● 包括全球信息服务，如文书网站、专家和从业人员网络、信息交流机制以及区域、次区域和/或国家级信息交换所机制网络。

● 包括样本获取和样本采集、技术获取和转让、能力建设、融资机会以及数据和知识分享等信息。

● 促进研究成果交流。

---

① 另见本节第 3.2 小节。

● 考虑小岛屿发展中国家和最不发达国家的特殊情况。

84. 可要求船旗国在交存材料后向信息交换所报告从国家管辖范围以外区域获取海洋遗传资源的情况。也可向公共保藏中心提供样本。

85. 为利用海洋遗传资源，需向信息交换所提供足够的资料。

86. 可为各种组织创建合作和协作平台，以更好地分享数据/信息。

6. 能力建设和海洋技术转让①

87. 海洋遗传资源活动的提案国需特别向小岛屿发展中国家提供能力建设。

88. 能力建设的要素可包括：

● 提供科学和技术以及政策和治理方面的教育/培训，包括通过设立全球奖学基金和开展联合研究，来加强在海洋遗传资源研究与开发方面的合作。

● 设立一个为小岛屿发展中国家提供专项拨款的基金。

D. 海洋保护区等划区管理工具措施

1. 海洋保护区等划区管理工具的目标②

89. 国家管辖范围以外区域海洋生物多样性的养护和可持续利用。

90. 海洋环境的保护和养护。

91. 保护、维护和恢复海洋健康，包括关键生态系统过程、栖息地和物种，以及易受影响的区域，包括易受气候变化影响

---

① 另见本节第 3.2 小节。

② 需要确定概览（二）是否需为划区管理工具（包括海洋保护区）提供一份排他性目标清单。

的区域，如独特、脆弱/敏感、稀有或高生物多样性的栖息地特征以及对特定种群或稀有或濒危海洋物种（如繁殖或产卵场）的生存、功能或恢复或对为大型生态系统提供支持至关重要的遗传区域。

92. 通过有效管理的全球性、连贯性和代表性国家管辖范围以外区域划区管理工具（包括海洋保护区）网络等，对海洋生态系统、生物多样性和栖息地进行保护。

2. 指导原则和方法

93. 可采用的指导原则和方法包括：

• 与《联合国海洋法公约》《联合国鱼类种群协定》和其他有关国际文书保持一致。

• 不损害现有相关法律文书和框架以及相关全球、区域和部门机构。

• 利用现有相关国际文书和框架以及相关全球区域和部门机构已开展的工作和专业知识。

• 尊重沿海国对大陆架（包括超过 200 海里以外的大陆架）的管辖权和权利。

• 在专属经济区和国家管辖范围以外区域采取措施的兼容性。

• 国际合作与协调。

• 必要性。

• 相称性。

• 生态系统方法。

• 预防原则/方法。

• 采用最佳科学信息。

• 一体化方法。

- 预防原则。
- 基于威胁的方法。
- 代表性。
- 适应性管理。
- 保护和保全海洋环境。
- 不同级别的保护。
- 根据具体情况，单独和临时性建立和管理划区管理工具（包括海洋保护区）。
- 可持续利用。
- 海洋环境保护和养护活动与其他海上合法活动之间的利益平衡。
- 多用途方法
- 公平利用。
- 透明度。
- 包容性。
- 公众参与。
- 咨询。
- 问责。
- 污染者付费原则。
- 可追溯性。
- 法律责任。
- 为当代和子孙后代管理海洋环境。
- 不对沿海国家造成过重负担。
- 当地土著居民和社区的传统知识。

3. 建立划区管理工具（包括海洋保护区）的程序

94. 方案 1："全球模式"——建立一个全球总体框架，来识别、指定、管理和执行国家管辖范围以外区域的划区管理工具，包括海洋保护区。

95. 方案2："混合模式"——在全球层面制定通用指南和目标，以加强合作与协调，并对区域和/或部门机制的决策和执行情况进行监督。

96. 方案3："区域和/或部门模式"——在全球层面提供总体政策指导，以促进合作和协调，同时认可区域和部门组织在决策中有充分的权力，且不受全球机制的监督。

97. 该进程并非一成不变。相反，它将考虑个案的具体的情况，同时考虑到全球文书机构制定的通用指南。

3.1 区域识别

98. 根据最佳可获科学信息，确定需通过划区管理工具（包括海洋保护区）进行保护的区域。

99. 区域确定准则包括：

● 基于现有国际公认标准（如关于《生物多样性公约》中具有重要生态或生物意义的海洋区域、海事组织指定的特别敏感海域、脆弱海洋生态系统和国际海底管理局指定的特别环境利益区的标准）的通用准则和/或指南，包括独特性、稀有性、脆弱性、易损性、敏感性、代表性、依赖性、自然性、生产力和多样性。

● 区域协定中的准则。

● 可适当考虑对全球代表性网络的贡献。

● 某些物种或生态系统与生计和粮食安全的相关性。

● 社会经济因素也是确定划区管理工具和海洋保护区规模和位置的关键。

● 根据从主管科学咨询机构获得的最佳科学信息和建议，

31

进一步制定准则。

- 各国和决策机构根据新国际文书商定的新准则。

100. 可利用国际海事组织已开展的工作、对具有重要生态或生物意义的海洋区域进行描述的《生物多样性公约》以及区域渔业管理组织，确定优先保护生物多样性区域和脆弱海洋生态系统。

3.2 指定程序

3.2.1 提案

101. 提案方可包括：

- 单独或集体或与有关组织合作的新国际文书的缔约国。
- 有权成为缔约方的国家和实体。
- 新国际文书下的科学/技术委员会。
- 在其职权范围内的其他政府间组织。
- 非政府组织。

102. 鼓励提案方在制定提案的过程中征询民间团体等利益攸关方的意见和建议。

103. 提案需考虑最佳可获科学以及预防方法/原则，并遵循生态系统方法。

104. 提案的内容可包括：

- 对区域/空间边界的划定。
- 保护目标描述。
- 描述该区域的特点和生物多样性价值以及有关物种/栖息地的敏感性，包括对海洋生态系统现状的评估。
- 对影响（包括累积影响）的描述、对威胁以及可能产生不利影响的活动的识别。
- 达到指定目标所需的养护或管理措施，包括被禁止的人类活动、管理计划和社会经济缓解措施等。

- 如有必要，拟议区域对具有生态代表性的海洋保护区网络所起的促进作用进行描述，包括对其与现有海洋保护区或其他划区管理工具之间的关系的描述。

- 关于毗邻区域的信息，包括国家管辖范围内的区域，如《联合国海洋法公约》第76条所覆盖的区域。

- 相关可能干扰海洋的其他合法用途的信息，并酌情考虑可能涉及的社会经济成本。

- 关于为实现养护目标而采取相关行动的国际组织和机构的信息，包括其他国际组织所采取的现有保护措施。

- 是否已与这些组织进行了事先磋商。

- 识别拟议划区管理工具与现有划区管理工具之间的重叠部分，包括协调措施。

- 时间安排：

Ø 方案1：为养护和管理措施设定一个时间段。

Ø 方案2：不设定养护和管理措施持续时间。

- 监测和审查，包括研究和监测计划要素。

- 执行计划。

3.2.2 提案咨询和评估

对提案的咨询和评估程序可包括：

105. 可通过秘书处把提案分发给：

- 方案1：缔约国和相关全球、区域或部门组织和框架。

- 方案2：所有国家，包括非缔约方和相关全球、区域或部门组织和民间团体。

106. 可邀请各国、相关全球、区域或部门组织和民间团体代表就该提案提出意见。

107. 在考虑建立划区管理工具时，需获得与国家管辖范围以外建立划区管理工具毗连区域沿海国的同意书。

108. 对提案反馈和评论意见设定一个时间期限提交。可在新国际文书中确定咨询时间期限。

109. 秘书处将向公众公开咨询过程中各方提出的意见和建议，并将其整理、汇编，然后转交给提案国。

110. 提案国将在必要时，根据咨询过程中收到的意见，修改其提案。

111. 可通过科学审议和建议机制（如科学/技术机构）等对提案进行审查，包括考虑是否存在类似的海洋保护区，以及该区域如何对新国际文书下的保护区进行补充，并就提案与文书中科学准则的兼容性提出意见和建议，包括就具有代表性的海洋保护区网络和生物地理分区方案提出建议。

112. 科学/技术机构可就该提案提出建议。

3.2.3 决策

在划区管理工具（包括海洋保护区）指定做出决定时，可考虑以下几点：

113. 划区管理工具（包括海洋保护区）的指定或设立需符合《联合国宪章》《联合国海洋法公约》和其他有关法律文书的宗旨和原则。

114. 新国际文书需要对现有区域和部门机构间的合作和协调情况进行规定。应允许相关现有框架和非缔约方作为观察员参加缔约方会议等全球机构的讨论。

115. 需考虑沿海国在其国家管辖范围内采取的海洋生物多样性养护和可持续利用措施。

116. 应特别考虑毗邻国家，以保证所采取的措施不会损害其可持续发展。

117. 在指定划区管理工具（包括海洋保护区）时，还应考虑气候变化和水下噪声的影响。

118. 各个海洋保护区的保护目标，以及其特征/生态系统的脆弱性及所受压力不同，所需的保护程度也不同。

119. 关于划区管理工具（包括海洋保护区）的时间期限：

● 方案1：指定无时间限制。

● 方案2：指定有时间限制。

● 如果在稍后阶段，海洋保护区完全落入沿海国拥有主权或管辖权的海域范围内，那么对其的指定将不再有效。沿海国可根据其国内法律采取类似措施。在存在部分重叠的情况下，相应修改海洋保护区的空间边界。

120. 需基于科学数据作出关于划区管理工具的决定，且划区管理工具具有普遍性和约束力。

121. 方案1：全球机构将作出以下决定：

● 指定海洋保护区的空间边界。

● 建立海洋保护区。

● 在海洋保护区中采取适当的保护和管理措施。

● 在作出决定时，需尽一切努力达成共识。可采用多数投票法。

● 国际海底管理局可作为一个重要的组成部分，因为其职权已得到了《联合国海洋法公约》的认可。

122. 方案2：把指定区域以及有可能对该区域造成不利影响活动的信息提交给对该活动有管辖权的机构和框架，以便其对该活动进行审议，并采取相关管理措施或其他行动。

● 当指定区域有多个可能对该区域造成不利影响活动有管

辖权的机构时，可建立这些机构的协调和合作程序，包括建立审议和酌情执行相关管理措施的程序。

• 如果现有框架决定采取与缔约方会议商定措施不同的措施或不采取措施，缔约方大会应与之磋商。新国际文书缔约国和有关现有框架应尽可能在相关框架内合作，以确保现有框架适当尊重缔约方大会的决定，并采取适当的养护和管理措施。

• 可考虑采取一些方法和手段，使有关措施对所有缔约国（包括有关现有框架的非缔约方）皆具有约束力。

123. 方案 3：可在现有国际机制中处理与划区管理工具（包括海洋保护区）的建立和管理有关的问题。

• 可在公开磋商后，向适当的区域海洋保护组织提出海洋保护区提案，该组织将根据新国际文书的要求以及公众咨询意见对提案进行审议。

• 区域海洋环境保护组织有权通过海洋保护区提案，包括采取任何属于其职权范围内的措施，并在其网站上公布其决定。区域海洋环境保护组织根据新国际文书的要求作出的决定对新国际文书所有缔约国皆具有约束力。

• 区域海洋环境组织或其缔约方可向国际海事组织、区域渔业管理组织和其他区域海洋环境组织，如国际海底管理局等，转交关于海洋保护区的决定。缔约国有义务在所参与的所有机制中努力实现新国际文书的目标。

• 其他相关机构应考虑其所管理的活动是否与养护目标相关，以及是否需在其职权范围内采取相关措施。

• 在作出关于海洋保护区的决定后，将召开新国际文书的缔约国会议，对问责制、透明度、审查和利益攸关方的参与进行规定。

124. 如果没有主管机构就拟议区域内某一活动的影响提出解决措施：

- 方案1：由缔约方确定实现该区域养护目标的具体措施。
- 方案2：由缔约方以及有权成为缔约方的国家和实体制定并审议具体措施。
- 方案3：新国际文书鼓励有关国家和组织建立新的组织或框架，并参与新组织或框架的活动。

125. 在指定海洋保护区之后，新国际文书下的秘书处将把新建立的海洋保护区以及其目标，管理措施以及监测和审查计划，通知给缔约国和非缔约方、相关全球、区域或部门组织和框架，以及所有其他利益攸关方。

4. 执行

126. 海洋保护区管理措施将在决策机构通过后若干天内生效，届时其将对缔约国管辖或控制下的进程和活动以及相关国际组织具有约束力。

127. 新国际文书缔约国应作为船旗国对其管辖或控制下的活动和进程进行管理。有关船旗国和港口国可在执行海洋保护区措施方面进行合作。

128. 如果执行一项海洋保护区管理措施将会阻止某个缔约国履行其在另一相关法律文书或框架下的现有义务，或阻止相关全球、区域或部门机构履行其现行义务，则管理措施不对该缔约国生效。如果此种情况不再存在，则海洋保护区管理措施将被视为对有关缔约国具有约束力。

129. 缔约国应承诺尽最大努力确保其所在主管区域性或全球性组织采取必要措施。

130. 可要求负责有关海域的主管国际组织实施和执行划区

管理工具（包括海洋保护区）。还可要求其采取必要措施来实现新国际文书的养护目标。同时要求其制定和实施生物多样性战略和行动计划，并在管理和决策过程中考虑海洋生物多样性养护。

131. 要确保在确定所要采取的划区管理措施时，考虑各国在国家管辖范围内采取的与海洋生物多样性养护和可持续利用有关的行动，以及毗邻国家管辖范围以外区域沿海国的利益，这点非常重要。

132. 缔约国有权对其船舶或受其控制和管辖的活动与进程，采取其他比主管国际组织的措施更为严格的措施。

133. 管理计划不适用于新国际文书的非缔约方，但应告知非缔约方海洋保护区的指定情况，并请其考虑在其管辖范围内可能对海洋保护区的养护目标产生影响的活动和进程，采取适当管理措施。

134. 养护措施和管理措施的实施需符合《联合国海洋法公约》的要求，包括但不限于主权豁免（第 236 条）和第 237 条中的义务条款。

5. 与现有措施的关系

135. 现有机制中关于划区管理工具（包括海洋保护区）的条款和新国际文书中此类条款之间的关系涉及现有国际条约与新国际文书之间的关系，应根据《维也纳条约法公约》中的条约适用一般原则予以处理。

136. 划区管理工具（包括海洋保护区）不应损害现有海洋保护区以及有关全球、区域和部门机构执行的相关条款。

137. 新国际文书中的任何内容皆不应损害各国，以及相关全球、区域和部门机构根据其他法律文书和框架指定划区管理工具（包括海洋保护区）的能力，也不应损害各国根据该类指定所负有的义务。

138. 关于区域和部门划区管理工具（包括海洋保护区）的认可程序：

● 方案1：现有区域和部门机制通过的划区管理工具（包括海洋保护区）需得到全球机制的认可。认可不应减损任何机构采取措施的权力。

● 方案2：部门划区管理工具（包括区域渔业管理组织指定的脆弱海洋生态系统、国际海事组织指定的特别敏感海域、国际海底管理局指定的特别环境利益区）不需得到正式的全球认可，但应通知全球机构，并将其纳入信息交换所机制和信息分享机制。

6. 能力建设和海洋技术转让

关于能力建设和海洋技术转让的条款包括：

139. 对发展中国家提供必要的支持，包括发达国家缔约方有义务在提案制定过程中提供技术、科学和资金支持，对帮助其对提案进行审查以及制定管理措施并对监测划区管理工具。

140. 避免将不成比例的养护负担向小岛屿发展中国家转移过重的养护负担（如《联合国鱼类种群协定》第7条）。

7. 监测和审查

关于监测和审查的条款包括：

141. 基于最佳可获科学并根据所确定的目标，对划区管理工具（包括海洋保护区）进行定期审查和监测，以评估其有效性。

142. 可根据审查结果，对某海洋保护区养护和管理计划以及具体措施进行调整，以反映该区域的状况。

143. 关于科学/技术机构的作用：

- 方案 1：新国际文书下的科学/技术机构将负责对划区管理工具（包括海洋保护区）进行监测和审查。
- 方案 2：在可能和适当的情况下，可把该职能授予区域机构。

144. 缔约国和主管全球、区域或部门组织需定期汇报其职权范围内活动措施的执行情况。为此，新国际文书可提供标准化的报告模板，并附报告时间表以及报告方式，如通过相关区域和部门机构进行的报告。

145. 可向秘书处提交报告，由其转交给主管科学咨询机构；该机构将对报告进行审议并提出建议，并酌情将其转交给缔约国供其审议和作出决定，同时公布报告。

146. 监测和审查进程应考虑根据新国际文书设立的科学委员会、各国、区域和部门机构以及相关全球和区域进程与框架（如海洋环境状况，包括社会经济方面，全球报告和评估常规进程）以及民间机构提供的科学数据和资料。也可对气候变化等外部因素进行说明。

147. 可在审查后发布进度报告，以查明缔约方、非缔约方和区域或全球机构的不足，这些不足会影响新国际文书所采取措施有效性。

148. 可在审查后对海洋保护区的现状进行维护或修改所制定的管理措施或取消该海洋保护区。在必要时，可基于最佳可获科学信息以及主管科学咨询机构的建议，调整保护目标和/或管理措施。

149. 尽量避免损害活动向其他地方转移的一种方法是通过全球管理机构对转移风险进行监测和评估，并采取相关措施。

150. 需根据新国际文书建立一个全球监测、控制和监督系统，以确保各保护区实现其目标，并查明船舶违规行为以及经常性违约行为。该机制将有助于实现现有监测、控制和监督系统之间的信息分享和协作。

151. 可根据新国际文书建立一个遵约机制。根据审查之后的结果，缔约方、利益攸关方，包括民间团体以及遵约委员会，可提交一份违约报告。当某一缔约方被确认未能履行新国际文书下应承担的义务，或非缔约方因未采取措施或行使有效控制权来确保其船舶或国民不从事任何破坏新国际文书中养护措施有效性的任何活动，而未能根据国际法的规定在海洋环境保护和养护方面进行合作，遵约委员会应就如何纠正其不当行为或不履约提出建议。应通知不遵约缔约方和非缔约方在指定的合理期限内对所声称的不遵约事项作出响应，并及时纠正其不当作为或不行为。在必要时，新国际文书将根据遵约委员会的建议采取相关措施促进各方遵约（如技术援助和能力建设）。如果缔约方或非缔约方继续损害保护区管理的有效性，和/或所保护的生态系统或其任何组成部分受到严重威胁，新国际文书的其他缔约方应采取适当的、旨在确保实现该区域养护目标的应对措施。

E. 环境影响评价

152. 在制定环境影响评价规定时，可参考国际公认的标准、进程和议定书，包括以下文书中的标准、进程和议定书：

● 《联合国海洋法公约》。

● 《跨界环境影响评价公约》（《埃斯波公约》）。

● 修订后的《生物多样性公约》之《海洋和沿海地区环境影响评价和战略性环境评价中考虑生物多样性时使用的自愿准

41

则》。

- 联合国粮食与农业组织的《公海深海渔业管理国际准则》。

- 国际海底管理局为指导承包商评估"区域"内海洋矿产资源勘探可能对环境产生影响的建议。

- 《南极条约环境保护议定书》。

1. 开展环境影响评价的义务

153. 根据《联合国海洋法公约》第 206 条的规定，各国需开展环境影响评价。

154. 可通过制定指南/指导方针等，对该义务进行规定。

- 科学/技术机构可制定并向全球决策机构提出关于指南，包括国家管辖范围以外区域拟议活动的环境影响评价标准和阈值的建议。

155. 开展环境影响评价的义务应由对所涉活动的发生地有管辖或控制权的国家，即对特定活动拥有有效控制权或对特定活动（而非简单地由悬挂某一缔约国国旗的船舶开展的活动）以许可或资助形式行使管辖权的国家承担。

156. 环境影响评价可由在国家指导和控制下的第三方，如研究机构或私人企业开展。

2. 指导原则和方法

157. 环境影响评价有助于实现对国家管辖范围以外区域海洋生物多样性养护和可持续利用。

158. 可采用的指导原则和方法包括：

- 预警原则/方法。

- 生态系统方法。

- 国际合作。

- 一体化方法。
- 使用最佳可获科学/基于科学的方法。
- 透明度。
- 包容性。
- 咨询。
- 公平。
- 有效性。
- 代内和代际公平。
- 保护和保全海洋环境的责任。
- 污染者付费原则。
- 管理工作。
- 零净损失原则。

3. 需开展环境影响评价的活动

159. 环境影响评价义务涉及各国管辖或控制下的拟议活动。

160. 确定需开展环境影响评价的可能方法包括：

- 方案 1：对于所有拟在国家管辖范围以外区域进行的活动，均需开展环境影响评价。
- 方案 2：在特定情况下需开展环境影响评价，包括基于：

Ø 可采用的阈值：

§方案 1：根据《联合国海洋法公约》第 206 条（"有合理依据认为"拟议活动会对环境造成严重污染或重大和有害变化）。

§方案 2：比《联合国海洋法公约》更严格的要求，包括"任何有害"变化。

§方案 3：把"轻微或短暂影响"作为判断是否需进行初步评估，来确定是否有可能产生重大有害影响，以及是否需启动

43

正式环境影响评价和报告的初始阈值。

§ 方案 4：高于"轻微或短暂影响"。

§ 为根据新国际文书所指定的使用划区管理工具的区域，或以其他方式指定的在国际层面上具有重要意义/脆弱性的区域（如具有重要生态或生物意义的海洋区域、脆弱海洋生态系统、特别敏感海域和海洋保护区），确定具体的阈值。

§ 可在附件中列出阈值。

Ø 活动清单包括：

§ 方案 1：制定一份需开展环境影响评价活动的指示性清单（参阅《埃斯波公约》附录 3）。

Ø 该清单不是穷尽，且不具有法律约束力。

Ø 可在附件中给出该清单。

§ 方案 2：制定一份免于环境影响评价活动的清单。

§ 该清单可由负责指导环境影响评价的主管机构制定。

§ 缔约国会议可对清单进行审查或更新，以反映新的和新兴活动以及科技发展。

Ø 如果一项国家管辖范围以外区域的活动已被现有义务和国际文书所涵盖，可采用的方法包括：

§ 方案 1：没必要为新国际文书下的活动进行额外的环境影响评价。

§ 方案 2：根据定义，应对此类活动开展环境影响评价。

• 对于根据新国际文书所指定的使用划区管理工具的区域，或以其他方式指定的在国际层面上具有重要意义/脆弱性的区域（如具有重要生态或生物意义的海洋区域、脆弱海洋生态系统、特别敏感海域和海洋保护区）内的活动，需开展环境影响评价。

161. 环境影响评价与全球进程（如海洋酸化和全球变暖）

无关，后者取决于许多因素，且目前由相关国际机构负责实施。

162. 应考虑累积影响：

- 方案1：考虑可能产生的累积影响，包括拟议项目可能会增强的气候变化、海洋酸化和低氧效应的累积影响。

- 方案2：应尽可能对累积影响进行评估。

4. 环境影响评价程序

163. 开展环境影响评价的一般程序性步骤包括：

- 筛选

Ø 方案1：由对拟议活动有管辖权的缔约国作出是否需开展环境影响评价的决定。

Ø 方案2：由新国际文书下的全球机构决定何时以及如何对国家管辖范围以外区域的活动开展环境影响评价。

- 范围界定

- 影响评估与评价

Ø 需对拟议活动的各个方面进行潜在影响评估。

Ø 应采用公认的科学方法为基础研究，拟议活动对海洋环境造成的风险和潜在影响或作用进行相关评估和分析。

- 环境影响评价报告

Ø 应根据《联合国海洋法公约》第205条和第206条的规定，公布和传播环境影响评价结果。

- 审查/监测

Ø 根据《联合国海洋法公约》第204条，各国需在得出环境影响评价的积极结果后，继续监测活动的影响。

164. 公众参与/介入的可能方法如下：

- 方案1：在从范围界定阶段开始的环境影响评价各个阶段进行参与和磋商。

- 方案 2：公示和磋商的类型与频率将根据活动的风险水平及所预期的影响确定。

- 方案 3：利益攸关方可在作出决定前提出建议。

165. 进行公众磋商的利益相关者可以为：

- 可能受到活动影响的相邻沿海国家。

- 新国际文书的缔约国。

- 活动所在地的区域或部门机构。

- 相关政府间组织和非政府组织。

- 科学界的相关专家和/或根据新国际文书设立的科学技术委员会。

- 受影响的行业。

166. 相关缔约国可向所有相关利益攸关方分发一份评估草案，其中包括公众意见、建议以及新国际文书所要求的信息。

167. 拟在国家管辖范围以外区域开展活动的发起者告知有管辖权的国家。

168. 科学/技术机构可：

- 监督环境影响评价过程。

- 审查提案，并就所提交的环境影响评价文件，包括一份针对国家管辖范围以外区域人类活动进行的累积影响评估，以及所拟议的关于环境管理计划的规定，包括监测、审查和遵约规定，向全球决策机构提供建议。

- 进行定期和事后评估。

- 可在科学/技术机构下设立能够对国家管辖范围以外区域的活动开展环境影响评价的专家组，并委托该专家组帮助那些能力不足的国家开展环境影响评价。

169. 战略环境影响评价/环境影响评价行政监督委员会可：

- 制定环境影响评价准则。

- 确保由适当的实体开展环境影响评价和战略环境影响评价，并向全球机构提供关于环境影响评价和战略环境影响评价的建议。

170. 如果国家管辖范围以外区域环境影响评价的权限属于特定部门和/或区域国际组织，各缔约国需直接或通过相关全球、区域或部门机构开展环境影响评价。

171. 将通过网站或其他方式公布评估草案以及随后的意见和建议。

172. 环境影响评价报告纳入缔约国会议报告。缔约国、有关机构、非政府组织等应有机会对评估文件、考虑事项和决定进行评估和审查。

173. 可由以下机构作出是否继续进行拟议活动的决定：

- 方案1：对所在区域活动有管辖和控制权的缔约国。

- 方案2：根据新国际文书设立的国际机构，如缔约方会议（根据该机构下的技术和科学委员会提供的建议，并遵照申诉程序）。

174. 须确保环境影响评价的结果，在活动授权、后续减缓或补偿（补救）措施以及减缓措施等级决定中得到充分考虑。

175. 只有在评估结果显示该活动不会产生重大不利影响或可设法避免此类影响的情况下，方可对拟议活动进行许可。应在作出活动许可决定时，要求提供一份环境管理计划。

176. 当活动不被授权时，可采用申诉程序。

177. 环境影响评价或缔约国基于环境影响评价作出的决定不接受任何外部实体或进程的审查。

178. 关于环境影响评价费用承担问题：

- 方案 1：新国际文书可指定环境影响评价费用的承担方。

Ø 环境影响评价费用可由运营商承担或由分担。

Ø 环境影响评价费用可由活动提案国承担。

Ø 发展中国家开展活动时，可考虑是否需进行筹资和/或其他合作手段（能力建设和技术转让）。

- 方案 2：把费用的决定权留给缔约国的国家权力机构。

5. 环境影响评价报告的内容

179. 环境影响评价报告的内容可包括：

- 对拟议活动及其目的的描述。

- 对可能受到环境影响（包括依存或相关生态系统、提供的生态系统服务、受影响的敏感或脆弱区域以及易受气候影响的脆弱区域）的描述。

- 对该区域所提供的生态系统服务的描述。

- 对拟议活动产生的潜在环境影响（包括对生态系统服务的影响）的描述。

- 累积、直接、间接、短期和长期的积极和消极影响。

- 在适当情况下，对拟议活动的合理替代方案，包括无行动替代方案的说明。

- 对旨在将不利环境影响降至最低的缓解措施的描述。

- 基线信息。

- 对预测方法和基本假设以及所使用的相关环境数据的说明；以及对在汇编所需信息时遇到的知识差距和不确定性的说明。

- 对环境影响评价的准确性和缓解措施的有效性进行验证跟进，包括一份项目监测和管理方案纲要，以及后续后分析计划（如适用）。

- 如有必要，制定一项修复计划。

- 执行和遵约条款。

- 非技术性概括。

180. 可制定一个通用的环境影响评价报告模板。

6. 跨界环境影响评价

181.《联合国海洋法公约》第 206 条可作为进行跨界环境影响评估的依据。

182. 不需为跨界环境影响评价制定单独程序。

183. 发生在国家管辖范围以内的活动可能而对国家管辖范围以外产生影响的情况，需考虑以下方面：

- 方案 1：所有可能对国家管辖范围以外区域产生重大不利影响的人类活动（不论其实际发生地在何处）都需开展环境影响评价。

- 方案 2：新国际文书不涵盖发生在国家管辖范围内活动开展的环境影响评价。

Ø 缔约国需对其国家管辖范围内的活动和报告公示，制定国家法律。

184. 发生在国家管辖范围以外区域对国家管辖范围以内区域或资源有潜在影响的活动，需进行跨界环境影响评价。

185. 当国家管辖范围以外的活动将对毗邻沿海国产生影响时：

- 方案 1：在项目规划和环境影响评价时，适当顾及沿海国，包括与沿海国进行磋商。

- 方案 2：通知沿海国密切参与环境影响评价过程，特别是评估工作；未经受影响沿海国的批准，不得开展相关活动。

- 向当地社区发出通知并向其咨询。

● 也可通知民间组织并向其咨询。

7. 战略环境影响评价①

186. 关于是否纳入战略环境影响评价相关规定：

● 方案 1：环境影响评价的对象是国家管辖或控制下的计划"活动"，不包括战略环境影响评价。

● 方案 2：对战略环境影响评价进行规定。

Ø 建立明确、透明和有效的战略环境影响评价要求和程序。

Ø 新国际文书中的环境影响评价的参数同样适用于战略环境影响评价。

Ø 促进各国就国家管辖范围以外区域的战略环境影响评价，通过特设机构或现有区域或全球部门机构，在区域层面进行合作。

§ 在需要开展环境影响评价的活动之前，在区域层面制定战略环境影响评价方案。

§ 鼓励区域和全球组织针对其职权范围内的区域制定战略环境影响评价方案。

§ 在新国际文书中制定关于国际海事组织和国际海底管理局等全球部门组织以及区域公约组织参与区域战略环境影响评价进程的机制。

Ø 战略环境影响评价应由集体资助。

8. 环境影响评价措施的兼容性②

187. 在比邻沿海国管辖范围的国家管辖范围以外区域开展环境影响评价时，考虑与沿海国措施的兼容性。

---

① 另见第 4 和第 12 小节。

② 另见第 6 节。

9. 与相关法律文书、框架以及全球、区域和部门机构环评之间的关系

与相关法律文书、框架及全球、区域和部门机构环评之间的关系包括：

188. 不应损害现有相关法律文书和框架，特别是《联合国海洋法公约》以及相关全球、区域和部门机构。

189. 尊重现有的人类活动对国家管辖范围以外区域生物多样性特征产生的影响进行环境评价的程序和指南，包括区域和部门机构制定的程序和指南。

190. 根据新国际文书进行的环境影响评价不应与相关现有机构进行的环境影响评价相重复。

191. 可允许继续进行区域和部门机构管理下的现有活动，但该类组织须在其活动管理过程中考虑人类活动影响环境。新国际文书可在协调以上方面发挥积极的作用。

192. 促进不同国际机构在环境影响评价方面的合作和信息分享。

193. 新国际文书下的全球机构应确保该进程的透明度和问责制的实施，并确保利益攸关方对评估和后续决策的审查权。

10. 信息交换所机制

194. 公开提供与国家管辖范围以外区域的环境影响评价过程相关的信息。

195. 可建立一个关于公共数据、环境影响评价和战略环境影响评价信息，以及关于国家管辖范围以外区域的基线数据的中央信息库，如信息交换所机制，包括：

- 发布环境影响评价草案。

- 允许利益相关者在规定期限内对环境影响评价草案发表

评论。

● 传播环境影响评价结果。

● 促进能力建设，特别是发展中国家的能力建设，包括获得分享全球最佳实践。

196. 通过联合国海洋事务和海洋法司管理的专业网站等，提高信息交换所的成本效益，并充分利用信息技术。

11. 能力建设和海洋技术转让

197. 需考虑发展中国家，特别是小岛屿发展中国家、最不发达国家和内陆发展中国家的特殊需要，包括必要的技术、知识和财政援助，基础设施和机构能力建设，以及海洋技术转让等。

198. 通过协作（如通过自愿同行审议机制或缔约国之间的"配对"），进行充分的能力建设和海洋技术转让。

199. 需充分考虑小岛屿发展中国家的特殊情况。

● 该机制需确保小岛屿发展中国家充分具备了解和理解环境影响评价内容及相关影响所需的能力。

● 为环境影响评价的开展和审查提供财政和技术支持，并鼓励提高在国家管辖范围以外区域开展活动过程中的公平性。

200. 发展中国家可根据具体情况，提交联合环境影响评价方案。

12. 监测和审查

可采用的监测和审查方法包括：

201. 根据《联合国海洋法公约》第 204 条，各国应对在环境影响评价后的活动影响，以及与其授权有关的条件（如预防、减轻或补偿措施）的遵守情况进行监测和评估。

202. 可按以下方式进行监测和审查：

● 方案 1：通过监督和审查机制，确保各方遵守新国际文书。

Ø 缔约国需每年编写并向审查委员会提交一份报告，详细说明对环境影响评价相关规定的执行情况。各国还可汇报其他缔约国未能执行环境影响评价相关规定的情况。报告应及时公布。

Ø 在秘书处和科学机构的协助下，委员会将编写一份关于各国遵守环境影响评价义务情况和违约事项的年度综合报告，并在随后公布该报告。

Ø 监督和遵约委员会应在活动监测和评估过程中，咨询受影响的沿海国家和相关区域/部门机构。

● 方案 2：由缔约国或活动决议国负责监测和审查工作，并定期向有关国家汇报。

203. 在活动结束后，为确保对环境的持续性保护，以自然资本核算的形式进行后续评估，并与筛选阶段确定的基准相比较。

204. 可设立一项应急基金，以减轻活动造成的潜在有害环境影响。根据污染者付费原则，活动拟议国将事先存入一笔商定的资金；该资金将在环境影响评价以及全球机构科学委员会批准后，返还给活动拟议国。

205. 公布后续措施和监测结果报告。

F. 能力建设与海洋技术转让①

206. 新国际文书可通过以下方式对旨在促进能力建设和海洋技术转让合作的一般义务进行规定。

---

① 可在概览（二）的各个部分或一独立的章节对海洋技术的能力建设和转让进行阐述。

- 方案 1：能力建设和海洋技术转让涉及各个方面，并影响一揽子方案中的所有其他要素，因此将其纳入新国际文书的其他章节。

- 方案 2：专门用一个章节来描述与其他章节有关联的各种要素。

207. 能力建设和技术转让条款应与《联合国海洋法公约》和其他国际协定中关于能力建设和海洋技术转让的条款协调一致，并应促进此类条款的执行或实施。

208. 新国际文书在制定条款时，可参考以下内容：

- 《联合国海洋法公约》第十四部分。

- 《生物多样性公约》第 18.1 条。

- 联合国教科文组织的《政府间海洋学委员会关于海洋技术转让的准则和指南》

- 《小岛屿发展中国家快速行动方式（萨摩耶路径）》（第 102、111 款）。

- 《关于持久性有机污染物的斯德哥尔摩公约》第 11 条和第 12 条的第 1 款和第 2 款。

- 《关于汞的水俣公约》第 14.1 条。

- 《伊斯坦布尔行动计划 2011—2020 十年期支援最不发达国家行动纲领》

1. 能力建设和海洋技术转让的目标

209. 新国际文书可就与国家管辖范围以外区域海洋生物多样性养护和可持续利用有关的能力建设与海洋技术转让的一般和具体目标作出规定。目标包括：

- 提高发展中国家在国家管辖范围以外区域海洋生物多样性的养护和可持续利用方面的能力。

- 提升、传播和分享与国家管辖范围以外区域海洋生物多样性的养护和可持续利用有关知识和专业技术，并授权所有国家充分参与实现新国际文书的目标。
- 协调海洋资源养护和可持续利用作业。
- 巩固和提高发展中国家实施新国际文书的能力。

2. 能力建设和海洋技术转让的指导原则和方法

210. 可采用的指导原则和方法包括：

- 根据《联合国海洋法公约》进行合作和协作的义务。
- 根据《联合国海洋法公约》促进各国海洋科技能力发展的义务。
- 根据《联合国海洋法公约》向发展中国家提供科技援助的义务。
- 基于最佳可获科学，提供数据和信息。
- 以需求为驱动和有意义。
- 长期支持。
- 针对性。
- 有效性。
- 平等。
- 互惠互利。
- 透明度。
- 综合方法。
- 根据《联合国海洋法公约》向发展中国家提供优惠待遇的责任。
- 特别注意发展中国家的需求（类似于《联合国鱼类种群协定》第七部分中的义务）。此项义务包括考虑有特殊利益和需求的国家：

Ø 小岛屿发展中国家。

Ø 最不发达国家。

Ø 内陆发展中国家。

Ø 地理不利的国家

Ø 非洲沿岸国家。

Ø 易受气候变化影响沿海国家。

Ø 中等收入国家。

● 促进妇女的作用和参与。

● 考虑《伊斯坦布尔 2011—2020 十年期支援最不发达国家行动纲领》中的原则。

● 涉及利益相关者，包括私营部门和组织。

● 《联合国海洋法公约》第十四部分为海洋技术的能力建设和转让提供了依据；《政府间海洋学委员会关于海洋技术转让的准则和指南》为能力建设和海洋技术转让提供了基础框架。

● 加强执行和借鉴现有文书和机制（包括《联合国海洋法公约》、国际海底管理局、政府间海洋学委员会和《气候变化框架公约》、《生物多样性公约》和《伊斯坦布尔 2011-2020 十年期支援最不发达国家行动纲领》）中汲取经验教训，并保证不损害或与之相重复。

● 优化现有财务、人力和技术资源的使用。

211. 关于知识产权与能力建设与海洋技术转让的关系：

● 方案 1. 确保对知识产权的保护。

● 方案 2. 适当考虑知识产权。

● 方案 3. 通过在主管机构，特别是主管与知识产权有关的世界知识产权组织或世贸组织贸易组织，查阅有关知识产权，力求在保护知识产权与促进和传播海洋技术之间取得平衡。

3. 能力建设和技术转让的范围

212. 能力建设和海洋技术转让可解决以下方面的问题：

• 实现对数据、样本、出版物和信息的访问、采集、分析和使用。

• 执行《联合国海洋法公约》规定的义务，以促进发展中国家的海洋科学研究能力建设和海洋科学技术转让。

• 从海洋科学相关活动中获益。

• 进行与获取和惠益分享有关的能力建设。①

• 开发、执行和监测划区管理工具（包括海洋保护区）。②

• 开展环境影响评价，并参与战略环境影响评价。③

• 实施可持续发展目标，特别是可持续发展目标 14。

3.1 能力建设

213. 关于应列入新国际文书的能力建设活动的类型，可考虑以下几点：

• 方案 1. 不提供列表，因为其可能有限制性，不能适应未来的变化。新国际文书可列出一般要求，详细信息将在稍后阶段由特设的工作组确定。

• 方案 2. 提供一个具有指示性、非详尽和灵活的活动清单。可包括以下内容：

Ø 通过各区域和部门机构在区域、次区域和国家层面采取措施执行新国际文书，发展人力资源和机构能力建设。

Ø 通过短期、中期和长期培训，奖学金和专家交流，进行个人能力建设。

---

① 另见第 C 节。

② 另见第 D 节。

③ 另见第 E 节。

Ø 通过设立全球奖学基金等提供科学、教育和技术援助，包括基础和应用在内的自然科学和社会科学，如海洋学、化学、海洋生物学、海洋地理空间分析、海洋经济学、国际关系、公共行政、政策和法律和科技培训。

Ø 协助制定、执行和实施国家法律、行政或政策措施，包括国家或区域层面的监管、科学和技术要求。

Ø 提升或加强有关组织/机构的能力。

Ø 访问和获取必要的知识和材料、信息和数据，以便为发展中国家提供决策所需的信息。

Ø 增强意识和知识分享，包括海洋科学研究信息。

Ø 制定联合研究合作计划、进行海洋科学技术和基础设施建设，并获取必要的设备，以维持和进一步发展国家的研发能力，包括数据管理能力。

Ø 科研项目和计划的协作与国际合作。

Ø 提升或加强有关组织/机构的能力。

3.2 技术转让

214. 关于海洋技术转让的定义应足够广泛，并考虑未来的科学发展。

215. 重点参考《政府间海洋学委员会关于海洋技术转让的准则和指南》，并考虑把技术转让所指的具体技术以及为满足新国际文书的要求可能对其作出的修正包括在内。

216. 技术转让可包括以下内容：

• 获得适当、可靠、价格合理、现代以及对环境友好的技术。

• 硬技术以及其他相关方面，如计算机、自主式水下潜器和遥控式水下潜器。

● 专业设备, 如声学和取样装置、多波束回声探测和声波水下定位系统等。

● 观测设施、设备以及原位和实验室观测设备, 如分析和实验设备, 以及对微生物至大型无脊椎动物进行高分辨率观察, 以进行海上和岸上 DNA 片段测序的分子工具。

● 进行高级数据分析和数据存储的 IT 基础设施, 包括高分辨率、大规模和长期数据采集设施。

● 数据和专业知识, 包括但不限于, 设备、手册、采样方法、标准、参考资料、准则、协议、样本、软件、方法和基础设施。

● 区域、次区域和国家层面的机构建设, 包括数据管理。

● 在非排他性基础上, 组装、维护和操作相关系统所需的培训、技术咨询和援助, 以及为此目的使用这些技术的合法权利。

● 海洋技术相关的创新融资机制。

4. 能力建设和技术转让的模式

217. 可对能力建设和技术转让作如下规定:

● 直接或通过适当的全球或区域组织和机构转让, 程序和模式必须清晰、简洁和有针对性, 并可能迅速地运作。

● 在逐案审议的基础上, 以国家特殊情况和具体需求驱动, 为所需国家提供量身定制的解决方案。

218. 能力建设和海洋技术转让需与国家和区域需求、优先事项和要求相一致, 并应适应不断变化的需求和优先事项。可基于以下内容对需求进行评估:

● 所有相关国家和利益攸关方在国家和区域层面对发展中国家的需求定期作出的评估。

- 对现有能力，包括机构和人力资源能力，作出的全面评估。
- 可持续发展目标的指标数据。

219. 可对海洋技术转让作如下规定：

- 方案 1：基于公平合理的条款和条件，以及有利的条款和条件。
- 方案 2：在自愿的基础上，基于互相商定的尊重知识产权以及促进科学、创新、研究和开发的条款和根据。
- 方案 3：在自愿的基础上，基于有利条款，包括共同商定的优惠条款和条件。

220. 进行各个级别的合作非常重要，促进方式包括：

- 进行南北合作、南南合作，以及与利益攸关方，包括政府间国际组织、非政府组织、学术界、商业/私营部门和慈善组织的三方合作。
- 区域海洋项目与区域渔业管理组织之间的合作。
- 与发展中国家的机构合作开展联合科研项目，并建立国家和区域科学中心，包括数据库。
- 进行知识分享，并提高国家管辖范围以外区域海洋生物多样性有效养护和可持续利用重要性的认识。
- 通过合资企业，促进人力资源开发、教育、技术援助/合作、技术开发和转让以及提供咨询服务。

221. 可通过以下方式实现人力资源开发以及与新国际文书目标和适用范围有关的技术和研究能力建设：

- 在国家、区域和全球各级提供培训机会，包括交流活动和研讨会。
- 提供指导并建立伙伴关系。

- 建立区域技术开发中心。

- 建立与联合国—"日本奖学金计划"类似的全球奖学金计划，以促进与国家管辖范围以外区域海洋生物多样性养护和可持续利用有关的科学、政策和治理研究。

- 建立汇集人力资源的强大的全球专业校友网络、促进联网和相互学习，并建立国际合作基金会。

222. 利用最佳实践以及从现有机制得出的经验教训，其中包括：

- 国际海底管理局下的机制。

- 《生物多样性公约》第 16 条。

- 《关于港口国预防、制止和消除非法、不报告和不管制捕鱼的措施协定》。

5. 信息交换所机制

223. 建立一个全球系统，将全球、区域和国家各级的信息交换机制网络连接起来，并提供中央"一站式"信息获取渠道。

224. 也可改善现有信息交换所机制之间的互操作性和联系。

225. 信息交换所机制的功能包括：

- 提供一个平台或存储库，包括作为一个集中信息访问点，用于传播、分享和协调知识，包括传统知识、数据和信息技术活动的集中信息访问点以及获取评估结果和出版物。

- 有助于确保快速/一站式地获取与新国际文书的目标和范围有关的能力建设和技术。

- 通过虚拟课程等方式，促进和便利获取相应专门知识和技术。

- 提供国家管辖范围以外的现有机会项目、活动和方案的信息，并把能力建设和海洋技术转让需求与机会相匹配。

- 确定最佳实践并识别差距，以更好地支持新国际文书的执行。
- 在国家、区域和全球层面开展行动。
- 促进国际协调与合作。
- 便利样本和知识的公开获取。

226. 秘书处或其他机构可负责信息交换所机制的管理。

227. 信息交换所机制可基于现有文书、机制和框架建立，但不应与之相重复，其中包括：

- 《联合国海洋法公约》第十四部分。
- 联合国教科文组织《政府间海洋学委员会关于海洋技术转让的标准和指南》、国际海洋学数据和信息交换，以及海洋生物地理信息系统。可对新国际文书与政府间海洋学委员会之间的关系进行说明，包括是否需要建立一个促进协调与合作的机构提供额外财政支持或资源，以强化该委员会的作用。
- 国际海洋管理局的文件。
- 《名古屋议定书》。
- 《联合国气候变化框架公约》和《巴黎协定》。
- 《粮食和农业植物遗传资源国际条约》全球信息系统。
- 《在环境问题上获得信息、公众参与、决策和诉诸法律的公约》（奥胡斯公约）。

6. 资金

228. 确保为相关能力建设和海洋技术转让提供充足、可预测和可持续的资金以及促进发展中国家私营部门与私人和公共实体之间建立真正的伙伴关系，可设立以下供资机制：

- 方案 1：设立一项自愿信托基金。
- 方案 2：利用现有供资机制，如全球环境基金。

- 方案3：设立一项特别基金以及其他独特的供资机制，如修复或责任基金以及应急基金等。
- 方案4：把自愿和强制性供资机制结合起来。

229. 资金来源包括：

- 自愿和强制性收益。利用现有筹资供资机制，如《名古屋议定书》和国际海底管理局的能力建设供资安排等。
- 为获取和利用海洋遗传资源所支付的资金、环境影响评价审批过程中支付的担保费、对不符合环境影响评价要求的罚金，以及一部分技术转让费用。
- 来自捐赠国或提出对国家管辖范围以外区域的海洋生物多样性资源进行勘探和开发的私营实体的资金，具体金额取决于所涉区域的面积、活动类型以及与拟议活动相关的风险等因素。

230. 把供资机制与气候变化机制和类似供资机制（如碳足迹）相结合。

231. 会员国、其他实体以及非政府组织、基金会、研究中心和个人等均可对该基金捐款。

232. 可考虑使用新的海洋可持续发展金融工具，如私人投资养护联盟。

233. 该基金可用于资助与能力建设和海洋技术转让有关的活动和项目，包括：

- 为发展中国家参加新国际文书下的重大会议提供资金。
- 协助发展中国家履行其在新国际文书下的承诺。
- 支持相关奖学金制度、项目、培训和其他机会，以帮助发展中国家的国民了解与国家管辖范围以外区域海洋生物多样性有关的活动并充分参与新国际文书的实施。

- 支持汇聚全球资源的区域科技中心，以加强技术转让工作。

- 支持为能力建设和海洋技术转让建立信息交换所。

234. 任何供资机制需尽可能减少资金获取和使用条件。

235. 应及时向目标国提供能力建设和技术转让资源。

236. 向小岛屿发展中国家和最不发达国家提供优先获得资金和其他优惠条件。

237. 该基金可专门用于易受气候变化影响的国家。

7. 监测、审查和后续行动

238. 监测、审查和后续进程：

- 应对发展中国家，特别是小岛屿发展中国家所面临的能力制约进行定期审查，以便在稳定和长期的基础上充分满足受援国和区域的需求。

- 基于在国家、区域和全球层面得到的定量和定性数据，衡量能力建设和技术转让工作。

239. 可通过以下机构或方式开展监测、审查和后续行动：

- 新国际文书下的咨询（科学和/或技术）或决策机构。

- 在秘书处和/或履约委员会支持下，定期举行缔约国审查会议和/或会议来评估需求，并弥补需求差距。

- 根据新国际文书，缔约国报告能力建设情况。

- 审查进程应包括对能力建设和海洋技术转让作出贡献的所有利益攸关方。

240. 确定对小岛屿发展中国家的报告要求（定期、透明、全面和精简），且该报告应有利于进行定期和系统性审查，包括对需求和优先事项的审查。

四 机制安排

241. 新国际文书可对机制安排作以下规定：

● 方案 1："全球模式"——在全球层面进行科学咨询、决策、审查和执行监测。

● 方案 2："混合模式"——在全球层面制定通用指南、准则和标准，而把科学咨询、执行和遵约方面留给区域和部门组织，同时在全球层面进行决策和执行情况监督。

● 方案 3："区域或部门办法"——建立一个致力于促进协调与合作的全球机制，同时把措施决定、跟进和落实审查权力留给区域和部门机构。

242. 可考虑采用现有机制的可能性。

A. 决策机构/议事机构

243. 关于新国际文书的机构设置，可建立一个定期召开缔约国会议的全球层面总框架。具体方式如下：

● 方案 1：建立一个定期召开缔约方会议的新国际机构。每年召开一次缔约方会议，每五年召开一次审查会议。①

● 方案 2：把对新国际文书执行情况的监督责任划归国际海底管理局。

---

① 另见第十一节。

244. 全球机构可由大会和理事会组成，成员数量有限，由大会选举产生。

245. 全球机构的职能可能包括：

- 根据最新可获科学信息，包括传统知识，制定目标、程序、准则、标准和指南。

- 监督/审查新国际文书的执行情况。

- 通过新国际文书执行有关的决定。①

- 设立附属机构，并在必要时向这些机构提供指导。

- 通过与现有机构建立合作与协调机制等，便利以及促进不同利益攸关方、国家和主管组织之间的合作与协调。

- 管理全球信息库。

- 审议并通过对新国际文书的修正。

- 促进与国家管辖范围以外区域海洋生物多样性养护和可持续利用有关的政策和措施间的协调。

- 确保遵守新国际文书的规定。

- 通过与新国际文书有关的工作计划和预算。

- 评估新国际文书在确保国家管辖范围以外区域海洋生物多样性的养护和可持续利用方面的有效性，并在必要时，提出加强新国际文书执行的措施，以便更好地解决海洋生物多样性养护和可持续利用方面出现的任何问题。

- 审查从缔约国和有关部门与区域机制收到的关于执行新国际文书所采取的措施的信息。

- 在缺少主管区域机构或该机构未能采取行动的情况下，通过关于如何执行新国际文书的决定。

---

① 另见第三（C）、（D）和（E）节。

● 审议缔约方所决定的其他问题。

246. 根据《联合国海洋法公约》第 208 条第 3 款的规定，全球机构所实施的法律规章和措施的效力"不得低于国际规则、标准以及建议的方法和程序"。

247. 非缔约方、有关政府间组织、非政府组织和其他利益攸关方均可以观察员的身份参加会议。

248. 在适当情况下，全球机构可征求现有区域和部门组织、民间机构和其他利益攸关方的意见。

249. 出于对发展中小岛国在某方面能力有限的考虑，全球机构会议应在大多数代表团，特别是发展中小岛国的常驻地举行。

250. 可在区域层面作出并执行决策，以充分反映区域和次区域的特点。

251. 鼓励缔约国尽可能通过区域文书进行合作，以执行新国际文书下通过的措施。

252. 在召开全球机构会议前，可建立区域/次区域议事机构并定期召开会议，以便：

● 根据全球准则、标准和措施，决定所要采取的措施。

● 与利益攸关方就相关项目进行广泛而全面的磋商。

● 向全球机构汇报。

● 向全球机构就新国际文书的执行提出改进建议或提交建议书。

253. 区域/次区域议事机构可由两部分（比邻沿岸国家和新国际文书的所有缔约方）组成；议事机构会议可向现有区域组织、现有部门组织、国际组织和其他利益攸关方的代表开放。

254. 如果存在次区域或区域组织或安排，且该组织或协会有能力制定养护和可持续利用措施，新国际文书的缔约国可能需要成为该组织的成员，以便与该组织或安排的其他成员有效合作，并积极参与该组织或安排的工作。

255. 如果某一缔约国不是次区域或区域组织或安排的成员，或未参与该次区域或区域组织或安排，该缔约国可根据有关国际协定和国际法，通过落实该组织或安排通过的养护和管理措施，开展有关渔业资源的养护和管理方面的合作。

256. 来自与国家管辖范围以外生物多样性有关的政府和非政府组织的代表，可作为观察员或以其他身份，按照有关组织或安排的程序，参加次区域和区域组织与安排召开的会议。以上代表可根据相关程序规定，及时查看这会议记录和报告。

B. 附属机构

257. 可设立一个作为附属机构的科学和/或技术机构，其中包括：

● 方案 1. 设立一个科学和/或技术机构。可利用现有框架下的科学委员会。其组织形式可为会或分委员会（与大陆架界限委员会类似）。

● 方案 2. 设立一个涵盖所有海域的科学委员会。

● 方案 3. 针对各个海域，设立多个科学委员会。

258. 以上机构的组成可包括：

● 由政府提名的多学科主题专家，包括来自新国际文书所涉问题的缔约国有关专家。

● 来自联合国粮农组织和国际海事组织等的代表以及专门从事新国际文书各项要素研究的国际专家。

● 传统知识专家或相关持有人。

259. 科学和/或技术附属机构的职能包括：

- 通过在原则上达成共识，作出决策。
- 向全球机构提出关于海洋遗传资源包括惠益分享问题①、划区管理工具包括海洋保护区②、环境影响评价③、能力建设和海洋技术转让的建议。④
- 识别与国家管辖范围以外区域海洋生物多样性养护和可持续利用有关的新问题。
- 就与国家管辖范围以外区域海洋生物多样性养护和可持续利用研究与开发方面的科学项目和国际合作提供咨询意见。
- 对决策机构及其附属机构提出的科学、技术和方法问题作出响应。
- 对所掌握的国家管辖范围以外区域海洋生物多样性的科学知识进行定期评估。适当考虑到可从一般进程以及其他相关进程（如描述具有重要生态或生物意义的海洋区域科学进程）获得的信息。
- 在必要时，执行财务、预算和法律等附加职能。

260. 可根据新国际文书设立以下附属机构：

- 战略环境影响评价/环境影响评价行政监督委员会。⑤
- 履约委员会，负责审查关于新国际文书的遵守和执行的一般性问题。⑥
- 财务和行政委员会。⑦

---

① 另见第三（C）节。
② 另见第三（D）节。
③ 另见第三（E）节。
④ 另见第三（F）节。
⑤ 另见第三（E）节。
⑥ 另见第七节。
⑦ 另见第六节。

- 能力建设和海洋技术转让委员会。
- 负责监督海洋遗传资源获取和惠益分享的机制/实体。①

261. 考虑到小岛屿发展中国家的特殊情况，每个附属机构可为小岛屿发展中国家分配专门的席位。

262. 为促进新国际文书的执行，可建立区域性机构，包括区域专家小组或委员会，如区域性管理委员会、区域能力建设和海洋技术转让委员会、区域执行委员会、区域财务和行政委员会。

C. 秘书处

263. 秘书处的职能可参考《联合国海洋法公约》第 319 条第 2 款和联合国大会决议第 49/28 号第 15 条以及《生物多样性公约》和《联合国气候变化框架公约》等其他文书中秘书处的一般职能确定。

264. 秘书处应在制定程序时考虑小岛屿发展中国家的特殊情况。

265. 不论是新设一个常设秘书处，还是由现有国际机构，如由联合国海洋事务与海洋法司提供秘书处服务，秘书处服务应以具有成本效益的方式提供。联合国海洋事务与海洋法司可作为新国际文书的秘书处，并配以必要的人力、技术和财政资源。

266. 秘书处不一定等同于受托人。

---

① 另见第三（C）节。

五　信息交换/交换所机制

267. 应促进国家之间以及相关区域、部门和国际组织之间的信息和数据交换（与《生物多样性公约》第 17 条类似）。

268. 应建立明确的信息原则，以便及时向缔约国、民间团体和外部机构提供会议文件、会议报告、决议、年度报告和本组织的监测结果。

269. 新国际文书可参考《南太平洋公海资源养护和管理公约》第 18 条，纳入关于透明度的条款，或在文书各个部分纳入促进/确保透明度的要求。

270. 各国、次区域或区域生物多样性管理组织和安排，应适当宣传养护和可持续管理措施，并确保有效传达这些执行措施所依据的法律、规章和其他法律规定。

271. 可建立信息交换所机制，以履行以下职能：

- 促进和推动信息、知识和数据分享。
- 促进和推动技术和科学合作。
- 维护缔约方和合作伙伴间的专家与从业人员网络。
- 采集以下信息：

Ø 海洋遗传资源以及有权访问所述资源的实体所存放的数据。

Ø 惠益分享可为货币和非货币形式，包括付款和财政资源。

71

Ø 关于划区管理工具和环境影响评价的科学数据，以及主管机构的后续报告和相关决定。

Ø 提供能力建设和海洋技术转让的机会。

• 与区域和国家信息交换机制相连。

272. 区域信息交换所机制将成为全球信息交换所机制的一部分。

273. 信息交换所机制可由秘书处管理。

274. 可采用一个逐步演变的方法，即秘书处在海洋遗传资源开采成为现实之前提供信息分享，在此之后建立一个专门机构。

六　财政资源和机制

A. 筹资机制

可采用的筹资方式包括：

275. 可通过以下方式获得执行新国际文书所需的资金：

• 强制性来源（缔约国的捐款、海洋遗传资源开采特许权使用费和里程碑付款）。

• 缔约国、非缔约国、国际金融机构、捐助机构、政府间组织、非政府组织，以及自然人和法人的自愿捐款。

276. 可设立一项全球信托基金，以履行以下职能：

• 资助发展中缔约国参与新国际文书的进程。

• 协助发展中国家履行其在新国际文书中的承诺，包括实施环境影响评价等。

• 为能力建设活动提供资金。

• 为与技术转让有关的活动和项目提供资金，包括培训。

• 支持地方社区传统知识的拥有者开展养护和可持续利用项目，包括在国家管辖范围内区域的项目，从而促进海洋管理的一致性。

• 支持国家和区域层面的公开咨询。

277. 由秘书处管理的捐赠基金可通过支持来自发展中国家的合格科学家和技术人员参与海洋科学研究项目和活动，以及为这些科学家提供参与相关方案的机会，促进和鼓励在国家管辖范围以外的区域开展合作海洋科学研究，包括与这些区域内海洋遗传资源有关的研究活动。

278. 对小岛屿发展中国家的特殊情况加以考虑的方法包括：

- 在基金中设定小岛屿发展中国家专项资金。
- 考虑为小岛屿发展中国家制定特殊程序（包括事前申请），通过触发支持机制用以协助准备所需的申请。
- 利用现有筹资机制。

279. 资金获取和报告程序应简单明了。

280. 财政与行政委员会可履行以下职能：

- 拟定财政细则、条例和程序草案。
- 评估缔约方的捐款。
- 拟定关于公平分享来自海洋遗传资源的资金和其他经济利益以及就此作出决定的规则、条例和程序草案。
- 促进为执行新国际文书目的而调动资源，并向缔约方提供援助，特别是发展中国家，其中又以最不发达国家和小岛屿发展中国家为重。
- 审查和编制预算。
- 监测新国际文书中设立基金。
- 向全球机构报告。

B. 修复/应急基金

解决修复和意外事件的方法包括：

281. 基于根据《联合国气候变化框架公约》和其他类似制度制定的"华沙损失和损害机制"所得出的经验，制定损失、损害和应急机制。

- 该机制为残余机制，即只有当主要负责人或责任人不能完全处理损害或修复时才被启用。

- 为资金（来自企业的预付款，赞助国家的债券保证金、自愿捐款以及强制性捐款等）确定明确的标准。

282. 可根据污染者付费原则，设立一项修复基金。希望参与国家管辖范围以外区域海洋生物多样性勘探和开采的私人实体将需根据活动可能对国家管辖范围以外区域海洋生物多样造成的环境损害程度为该基金捐款。如对国家管辖范围或其所在的国家管辖范围以外区域海洋生物多样性造成污染或其他破坏性影响，则该基金将被用于资助国家管辖范围以外区域海洋生物多样性的修复，包括对自然环境的修复。

283. 可设立一项应对环境灾害的应急基金，如人类活动造成的污染和其他毁灭性灾害。

七　执行

284. 可要求各国和所有参与生物多样性管理的实体针对新国际文书下的区域采取统一的措施，以实现对生物多样性的长期养护和可持续利用。

285. 各国（包括部门和区域组织，如有）应单独或集体负责新国际文书的实施，并确保悬挂其国旗的船舶、国民及其管辖下的实体遵守并执行新国际文书。

286. 可要求各国颁布法律、条例和/或采取必要措施，以确保遵守新国际文书中所规定的标准、措施和程序。

287. 可采用的履约监测和审查方法包括：

- 为国家管辖范围以外区域建立一个全球监测系统，以促进与现有监测系统之间的信息分享和协作。

- 可采用的机构设置包括：

Ø 方案 1. 由全球机构负责监测和审查各缔约国遵守新国际文书的情况，并在出现违反新国际文书相关条款的情况下，强制执行该条款。

§ 是否遵守新国际文书将是缔约国会议/会议所设立的具体被授权机构的审查重点。具体设立情况如下：

Ø 该委员会可设立一个向缔约国提供咨询和援助的分支机构，以促进遵约，并设立一个确定缔约国不履行其承诺的后果的执行部门。

Ø 应通知不遵约的缔约国和不合作的非缔约国，在指定的合理期限内对所提出的不遵约事项作出响应，并纠正其作为或不行为。

Ø 该机构也可接收来自非国家实体的违约投诉，然后进行进一步分析，并提请全球机构适当跟进。

Ø 方案 2. 由区域和部门机构负责监测和审查对新国际文书的遵守情况。

288. 建立把区域机构考虑在内的执行和实施机制。

289. 可建立定期报告和审查程序，以便：

● 缔约方和相关区域或全球机构定期报告养护和管理措施的执行情况。这些报告应予以公开。

● 提供科学委员会、所有相关区域或全球机构以及利益攸关方，包括民间机构的意见，以及通过全球监测、控制和监督系统收集的信息进行审查。

● 应在审查程序中公布进度报告，并查明缔约国、非缔约国和区域或全球机构所存在的、影响根据新国际文书所采取措施有效性的不足之处。

八 争端解决

可采用的争端解决的方法包括：

290. 设立一个争端预防机制。可由具体的委员会或选定的专家对这些问题进行争端。

291. 缔约国应通过和平途径解决一切与新国际文书的阐释与应用有关的争端。

292. 各缔约国可根据明确的双方协定，将争议提交给第三方解决。

293. 可考虑采用《联合国宪章》第 33 条中的方法。也可参考 1982 年《关于和平解决国际争端的马尼拉宣言》。

294. 关于《联合国海洋法公约》中与和平解决争端相关的规定：

- 方案 1：在根据新国际文书解决争端时，可先行参考这些规定。

- 方案 2：不适宜直接适用这些规定。

295. 可参考《联合国鱼类种群协定》中关于解决争端的程序。

296. 可授予国际海洋法法庭解决争端的司法权以及咨询权。可设立一个处理国家管辖范围以外区域海洋生物多样性问题的特别分庭。

297. 可设立一个类似于国际海洋法法庭的新机构。

298. 也可考虑区域争议解决机制。

299. 可考虑列入与《南太平洋公海渔业资源养护和管理组织公约》类似的退出机制；缔约国可选择一种退出机制，通过该机制，这样可继续推进相关措施，同时，也可为有关国家提供争端解决该事项的仲裁机会。

九　非缔约国

300. 鼓励非缔约国成为新国际文书的缔约国。

301. 并不免除新国际文书的非缔约国在《联合国海洋法公约》和习惯国际法中的一般性义务，包括保护和保全海洋环境、善意合作以及确保其活动不会破坏新国际文书中养护措施有效性的义务。

302.《维也纳条约法公约》中的有关规则也将适用。

十 责任与赔偿责任

303. 关于责任与义务的规定，可采用的方法包括：

- 方案1：列入一项与《联合国鱼类种群协定》第 35 条类似的规定。

- 方案2：列入一项反映并借鉴国际法中所规定的，缔约国有责任避免对国家管辖范围以外区域或其他国家造成损害，"确保在其管辖或控制范围内的活动不对其他国家或超出国家管辖范围的区域的环境造成损害"。

- 方案3：不在新国际文书中列入关于责任的条款，因为国际法委员会所拟订的并附于联合国大会第 56/83 号决议之后的"国家对国际不法行为的责任条款"是国际法在此领域的权威条款。

304. 可根据污染者付费原则、《国际法委员会关于危险活动跨境损害的条款》以及常规责任解决制度，制定指南。

305. 在确定责任范围时，可参考国际海底管理局的《"区域"内矿产资源开发规章草案》第 21 节。

306. 可参考 2011 年《国家担保个人和实体在"区域"内活动的责任和义务的咨询意见》。

十一　审查

307. 可设立一个类似《联合国鱼类种群协定》第36条规定的审查机制，对新国际文书的有效性和执行情况进行定期审查。应根据商定标准，在新国际文书生效后一段规定时间内（如五年后）进行审查，并在其后进行定期审查。

308. 需确保较短的审查间隔。

309. 通过审查，可了解以下方面：

- 根据新国际文书设立的机构履行其指定职能的情况。
- 根据新国际文书中的目标、原则和标准所作出的决定。
- 新国际文书的缔约国执行新国际文书的情况。
- 对国家管辖范围以外区域海洋生物多样性可持续利用负有责任的区域和部门机构履行其在新国际文书下的职能的情况。

十二　最终条款

310. 新国际文书将列入标准的最终条款，如《联合国鱼类种群协定》第37—50条和《联合国海洋法公约》第309—319条所载的规定，包括与争议解决、签署、批准、加入、生效、保留与例外、声明与陈述、修正、退出、国际组织参与、存管和正文有关的规定。

311. 需考虑新国际文书生效所需的批准书数量，以确保其迅速生效。

312. 关于参与：

● 寻求普遍参与。根据《联合国鱼类种群协定》第37—39条的规定，所有国家和其他实体均可签署、批准和加入新国际文书。

● 与《联合国海洋法公约》附件9中第305条的规定类似，新国际文书也可由国际组织签署，以允许欧盟参与。

● 可考虑新国际文书临时适用的必要性和可能性。

Chair's streamlined non-paper on elements of a draft text of an international legally-binding instrument under the United Nations Convention on the Law of the Sea on the conservation and sustainable use of marine biological diversity of areas beyond national jurisdiction

Explanatory note:

As indicated at the third session of the Preparatory Committee established by resolution 69/292: Development of an international legally binding instrument under the United Nations Convention on the Law of the Sea on the conservation and sustainable use of marine biological diversity of areas beyond national jurisdiction, this Chair's non -paper aims to provide a streamlined version of the Chair's non-paper issued on 28 February 2017. It is based on that non-paper and its supplement, and also takes into account the proposals, issues and ideas presented by delegations.

The present non—paper provides a reference document to assist delegations in their consideration of the issues addressed by the Preparatory Committee. It provides a compilation of the ideas and proposals put forward by delegations without consideration for the level of support for these ideas and proposals. The inclusion of ideas and proposals in this document does not imply agreement to, or convergence of views on, such ideas and proposals among delegations. Where options are presented, the order of such options should not be construed as indicating a suggested order of priority.

The content of this document is without prejudice to the position of any delegation on any of the matters referred to therein. Further, the elements listed are not necessarily exhaustive and do not preclude consideration of matters not included in this document.

The Chair wishes to express his appreciation to the delegations that made available to him their suggestions, proposals and position papers for the preparation of this non—paper. ①

---

① Available at http: //www. un. org/depts/los/biodiversity/prepcom_ files/rolling_ comp/Submissions_ StreamlinedNP. pdf

Ⅰ. PREAMBULAR ELEMENTS

1. Description of broader contextual issues, such as:

• The importance of marine biodiversity for ocean health, productivity, and resilience, food security, ecosystem services and sustainable development for present and future generations.

• The importance of both the conservation and sustainable use of marine biological diversity of areas beyond national jurisdiction.

• The link between climate change and oceans.

• The usefulness of environmental impact assessments (EIAs) for the prevention and identification of possible threats to the marine environment.

• The importance of the utilization of area-based management tools (ABMTs), such as marine protected areas (MPAs), to effectively protect and preserve marine biodiversity.

• The importance of capacity building and technology transfer for States, particularly for developing countries.

2. Description of the reasons for action, such as:

• The need for a comprehensive global regime to better address the conservation and sustainable use of marine biological diversity of areas beyond national jurisdiction.

- The need for close cooperation and coordination with the relevant existing bodies.

- The desire for an effective regime for the conservation and sustainable use of marine biological diversity of areas beyond national jurisdiction, including through a fair and equitable regime of access to and sharing of benefits of marine genetic resources.

- The importance of legal certainty in the regime for access to marine genetic resources, and benefit sharing.

3. Recognition that the United Nations Convention on the Law of the Sea (UNCLOS) sets out the legal framework within which all activities in the oceans and seas must be carried out.

4. Reference to, and recognition of the work under, relevant international instruments, such as the United Nations Fish Stocks Agreement (UNFSA), and bodies, such as the International Maritime Organization (IMO), the International Seabed Authority (ISA), regional fisheries management organizations (RFMOs), and regional seas organizations.

5. Reaffirmation of the jurisdiction and sovereign rights of coastal States over their continental shelf, including beyond 200 nautical miles, where applicable.

6. Some approaches and principles could also be set out in the preamble of the instrument, such as:

- The instrument should not undermine existing relevant legal instruments and frameworks and relevant global, regional and sectoral bodies.

- Common heritage of mankind.

87

- Freedom of the high seas.

- Special considerations for developing countries, particularly small island developing States (SIDS) and least developed countries (LDCs).

- Importance of global and regional cooperation.

- Fair and equitable participation of States in benefits derived from access to marine genetic resources.

- Common concern of humankind.

Ⅱ. GENERAL ELEMENTS

A. *USE OF TERMS*[1]

7. Definitions would need to be consistent with those contained in UNCLOS, UNFSA, and the Convention on Biological Diversity (CBD), including its Nagoya Protocol on Access to Genetic Resources and the Fair and Equitable Sharing of Benefits Arising from their Utilisation to the CBD (Nagoya Protocol), and other relevant international instruments, and adjusted to the context of marine biodiversity of areas beyond national jurisdiction.

8. The definitions would not be intended to cover trade in commodities.

9. Possible terms and definitions could include:

● Areas beyond national jurisdiction

○ "Areas beyond national jurisdiction" means the high seas and the Area, as defined in UNCLOS.

● Area−based management tools

○ *Option* 1: The definition of ABMTs could include three ele-

---

[1] Definitions could be included under the respective parts of the instrument, unless these terms were to be used in more than one part of the instrument.

ments：（1）the objective—ABMTs would be aimed at the conservation and sustainable use of marine biological diversity；（2）the geographic scope—ABMTs would be applied only to areas in the high seas and the international seabed area；（3）the function – ABMTs would include different functions and management approaches.

○ *Option* 2：ABMTs are tools designed and applicable in a specified area located beyond national jurisdiction with a view to achieving a defined objective（environmental conservation or/and resource management）.

○ *Option* 3：A spatial management tool for a geographically defined area through which one or several sectors/activities are managed with the aim of achieving particular objectives and affording higher protection than the surrounding areas.

○ *Option* 4：ABMTs include both sectoral and cross – sectoral measures that contribute to conservation and sustainable use of marine biodiversity. Cross – sectoral ABMTs, including MPAs, and marine spatial planning, are those tools that require cooperation and coordination across multiple organizations and bodies, may achieve broader objectives and respond to cumulative impacts. Sectoral ABMTs include measures adopted by a competent international organization to achieve biodiversity conservation objectives for a specific area and include fisheries closures designated by RFMOs, Particularly Sensitive Sea Areas（PSSAs）designated by the IMO, or Areas of Particular Environmental Interest（APEIs/reference zones）designated by the ISA.

● Biological diversity

○ "Biological diversity" means the variability among living organisms from all sources including, inter alia, terrestrial, marine and other aquatic ecosystems and the ecological complexes of which they are part; this includes diversity within species, between species and of ecosystems.

- Biological resources

○ "Biological resources" includes genetic resources, organisms or parts thereof, populations, or any other biotic component of ecosystems with actual or potential use or value for humanity.

- Bioprospecting
- Biotechnology

○ "Biotechnology" means any technological application that uses marine biological systems, living organisms or derivatives thereof, to make or modify products or processes for specific use.

- Continental shelf, as defined in UNCLOS
- Derivatives

○ "Derivative" means a naturally occurring biochemical compound resulting from the genetic expression or metabolism of biological or genetic resources, even if it does not contain functional units of heredity (based on Nagoya Protocol, article 2).

- Ecosystem

○ "Ecosystem" means a dynamic complex of plant, animal and micro-organism communities and their non-living environment interacting as a functional unit.

- Ecosystem-based management

○ "Ecosystem - based management" means an integrated approach to management that considers the entire ecosystem,

91

including all stakeholders and their activities, and resulting stressors and pressures with direct or indirect effects on the ecosystem under consideration. The goal of ecosystem-based management is to maintain or rebuild an ecosystem to a healthy, productive and resilient condition, through, inter alia, the development and implementation of cross-sectoral ecosystem-level management plans".

- Environmental impact assessment

○ "Environmental impact assessment" means a process to evaluate the environmental impacts of activity to be carried out in areas beyond national jurisdiction, with an effect on areas within or beyond national jurisdiction, taking into account interrelated socioeconomic, cultural and human health impacts, both beneficial and adverse.

- *Ex situ* collection
- Genetic material

○ *Option* 1: "Genetic material" means any material of plant, animal, microbial or other origin containing functional units of heredity (based on CBD, article 2).

○ *Option* 2: "Genetic material" means any material of plant origin, including reproductive and vegetative propagating material, containing functional units of heredity [based on International Treaty on Plant Genetic Resources for Food and Agriculture (ITPGRFA), article 2].

○ *Option* 3: "Genetic material" means any material of plant, animal, or microbial origin containing functional units of heredity collected from the Area; it does not include materials made from material, such as derivatives, or information describing material, such as genetic sequence data.

- (Marine) genetic resources

○ Definition must take into account the distinction between fish used for its genetic properties and fish as a commodity. The same distinction is relevant with respect to other animal species such as molluscs that can be used as commodities.

○ The definition could include the following elements: (1) animal, plant, microbe or other origin in the oceans and seas; (2) genetic materials containing functional units of heredity; (3) the actual or potential value; (4) the resources derived from areas beyond national jurisdiction.

○ *Option* 1: "Genetic resources" means genetic material of actual or potential value.

○ *Option* 2: "Marine genetic resources" means any marine genetic material of plant, animal, or microbial origin of actual or potential value collected from the Area.

○ *Option* 3: "Marine genetic resources" means any marine genetic material of plant, animal, microbial or other origin, containing functional units of heredity, being of actual or potential value.

- *In silico* access

- *In situ* collection

○ "*In situ* collection" means the collection of marine genetic material in ecosystems and natural habitats in areas beyond national jurisdiction.

- Marine protected areas

○ The definition could distinguish MPAs as a subcategory of

93

ABMTs which have a primary stated objective of achieving long-term conservation of marine biodiversity and ecosystems.

○ Any definition must be wide or flexible enough to encompass the high seas protected areas already created by RFMOs, so that they would be fully recognised as MPAs under the instrument.

○ *Option* 1: "Marine protected area" means a geographically defined area which is designated, regulated and managed to achieve specific conservation objectives (CBD, article 2).

○ *Option* 2: The definition of "protected area" provided in article 2 of the CBD is a starting point which would need to be adapted, as appropriate, in order to specifically focus on marine areas beyond national jurisdiction.

○ *Option* 3: "Marine protected area" means a designated geographically defined marine area [in areas beyond national jurisdiction] where human activities are regulated, managed or prohibited in order to achieve specific conservation objectives including the long-term conservation and resilience of nature.

○ *Option* 4: "Marine protected area" means a defined area of the marine environment, including its associated flora, fauna, historical and cultural features, which has been reserved by legislation or other effective means, including custom, with the effect that its marine biodiversity enjoys a higher level of protection than its surroundings.

○ *Option* 5: "Any marine geographical area that is afforded greater protection than the surrounding waters for biodiversity conservation or fisheries management purposes." MPAs would not be limited to marine reserved areas or no-take zones.

- Marine scientific research
- Marine spatial planning

○ Marine spatial planning is a cross - sectoral ABMT that provides a framework for the orderly and sustainable use of the oceans as envisioned by UNCLOS with a view to balance demands for development with the need to protect the marine environment. Sectoral AB-MTs (e. g. fisheries closures, PSSAs, APEIs), other cross-sectoral ABMTs (e. g. MPAs), strategic environmental assessments (SEAs) and EIAs are an integral part of this overarching planning approach. Marine spatial planning approaches would be ecosystem - based, adaptive and include all relevant stakeholders in the area under consideration.

- Sustainable use

○ "Sustainable use" means the use of components of marine biodiversity in a way and at a rate that does not lead to the long term decline of biological diversity, thereby maintaining its potential to meet the needs and aspirations of present and future generations.

- Technology

○ "Technology" means hard technology as well as all of its associated aspects, such as specialized equipment and technical know - how, including manuals, designs, operating instructions, training and technical advice and assistance, necessary to assemble, maintain and operate a viable system and the legal right to use these items for that purpose on a non-exclusive basis. It also refers to infrastructure and enhancing technical capacity to make such transfer sustainable.

- Transboundary environmental assessment
- Transfer of marine technology

○ The transfer of marine technology refers to the transfer of instruments, equipment, vessels, processes and methodologies required to produce and use knowledge to improve the study and understanding of the nature and resources of the oceans.

- Utilization of marine genetic resources

○ *Option* 1: "Utilization of marine genetic resources" means to conduct research and development on the genetic and/or biochemical composition of marine genetic resources, including through the application of biotechnology as defined in article 2 of the CBD.

○ *Option* 2: "Utilization of marine genetic resources" means to conduct commercial research and development on the genetic and/or biochemical composition of genetic resources, including through the application of biotechnology.

B. *SCOPE / APPLICATION*

1. *Geographical scope*

10. *Option* 1: Areas beyond national jurisdiction.

11. *Option* 2: Areas not adequately addressed by existing international conventions.

12. Does not apply to maritime zones under national jurisdiction, including the continental shelf beyond 200 nautical miles where applicable.

2. *Material scope*

13. All elements of the package specified in General Assembly resolution 69/292.

14. With regard to activities:

- *Option* 1: Conservation, sustainable use and responsible management of all marine living organisms of areas beyond national jurisdiction.

- *Option* 2: Activities carried out under the jurisdiction or control of a contracting party in areas beyond national jurisdiction.

- *Option* 3: Any activity or development that has the potential to impact on marine biological diversity of areas beyond national jurisdiction, including on ocean processes.

- *Option* 4: Activities with the potential to have significant effects on or to cause damage to marine biodiversity or ecosystems in areas beyond national jurisdiction regardless of where these activities occur.

- *Option* 5: All activities that take place in areas beyond national jurisdiction and/or may have an impact on marine biological biodiversity and resources of areas beyond national jurisdiction. Where such activities are already managed or governed by an existing agreement, the instrument would apply relevant provisions of the existing agreement *mutatis mutandis*.

- *Option* 6: All existing and new activities and sectors impacting on marine biodiversity of areas beyond national jurisdiction with respect to the elements identified in the "package", while not undermining existing relevant legal instruments and frameworks and relevant global, regional and sectoral bodies.

- *Option* 7: Activities not adequately addressed by existing international conventions, e. g., UNCLOS and CBD.

- *Option* 8：Fisheries management in areas beyond national jurisdiction would not form part of the negotiations.

3. *Personal scope*

15. The instrument would extend to States and entities in a manner similar to UNFSA.

C. *OBJECTIVE（S）*

16. Ensure the conservation and sustainable use of marine biodiversity of areas beyond national jurisdiction.

17. Possible additional objectives could include：

- Protect and preserve the marine environment.

- Furthering of regional cooperation and regional cooperative mechanisms.

- Prevent or eliminate excess capacity and ensure that levels of effort by entities involved are commensurate with the sustainable use of biological diversity as a means of ensuring the effectiveness of conservation and sustainable management measures.

D. *RELATIONSHIP TO UNCLOS AND OTHER INSTRUMENTS AND FRAMEWORKS AND RELEVANT GLOBAL, REGIONAL AND SECTORAL BODIES*

18. Relationship to UNCLOS：

- Nothing in the instrument would prejudice the rights, jurisdiction and duties of States under UNCLOS. The instrument would be interpreted and applied in the context of and in a manner consistent with UNCLOS.

19. Relationship to other instruments：

- The instrument should not undermine existing relevant legal instruments and frameworks and relevant global, regional and sectoral bodies.

○ *Option* 1: A without – prejudice clause would assist in achieving this purpose, including a provision similar to article 44 of the UNFSA to the effect that there would be no alteration of the rights and obligations emanating from other treaties.

○ *Option* 2: The instrument would not affect the competence of relevant international organizations and arrangements within their areas of competence.

• State explicitly what role or function the instrument would not have in relation to activities in areas beyond national jurisdiction, such as is done under the Convention for the Protection of the Marine Environment of the North–East Atlantic.

○ *Option* 3: The regulations and measures put in place by the instrument and the governing body "shall be no less effective than international rules, standards and recommended practices and procedures", in line with UNCLOS Article 208, para. 3.

○ *Option* 4: Standards applied in areas beyond national jurisdiction would not be lower than those for exclusive economic zones.

20. The provisions of the instrument would not apply to vessels entitled to sovereign immunity (in line with article 236 of UNCLOS).

21. Matters not regulated by UNCLOS or this instrument continue to be governed by the rules and principles of general international law.

Ⅲ. CONSERVATION AND SUSTAINABLE USE OF MARINE BIODIVERSITY OF AREAS BEYOND NATIONAL JURISDICTION

A. *GENERAL PRINCIPLES AND APPROACHES*①

22. A distinction could be drawn between principles and approaches.

23. Any definitions and/or interpretation of guiding approaches and principles would need to be consistent with those already agreed under UNCLOS, UNFSA, CBD and other relevant international instruments, such as the Rio Declaration on Environment and Development.

24. Possible principles and approaches could include:

- Recognition of need for a comprehensive global regime to

---

① Possible ways of referring to approaches and principles could include: 1) explicit reference to these approaches and principles in the instrument (in the preamble of the instrument, in a stand-alone article, or some approaches and principles may benefit from further elaboration in an article of their own, similar to Articles 6 and 7 of UNFSA); 2) reflecting these approaches and principles in the content of individual provisions of the instrument by making them operational.

better address the conservation and sustainable use of marine biological diversity of areas beyond national jurisdiction.

- Respect for the freedoms and the balance of rights, obligations and interests enshrined in UNCLOS.

- Incorporation of, and non – derogation from, the relevant principles enshrined in UNCLOS.

- Common heritage of mankind.

- Freedom of the high seas.

- Recognition of existing relevant legal instruments and frameworks and relevant global, regional and sectoral bodies (in particular UNCLOS, UNFSA, RFMO/As, IMO, ISA, and regional seas conventions).

- No undermining of existing relevant legal instruments and frameworks and relevant global, regional and sectoral bodies.

- Due regard for the rights of others.

- Respect for the rights of coastal States over all areas under their national jurisdiction, including their continental shelves beyond 200 nautical miles where applicable.

- Respect for the sovereignty and territorial integrity of coastal States.

- Compatibility.

- Adjacency and requirement to consult adjacent States.

- Recognition of the role of adjacent coastal States as well as other States.

- Enhanced cooperation and coordination between and among States and organizations to conserve and sustainably use marine biodiversity in areas beyond national jurisdiction.

- Protection and preservation of the marine environment and its biodiversity, including for the benefit of future generations.

- Duty not to transform one type of pollution into another or not to transfer damage or hazards.

- Use of biodiversity of areas beyond national jurisdiction for peaceful purposes only.

- Integrated approach.

- Ecosystem approach.

- Science-based approach.

- Use of the best available scientific information.

- Public availability of information.

- Public participation.

- Stakeholder involvement.

- Good governance.

- Transparency.

- Incorporation of traditional and local knowledge.

- Accountability.

- Equity.

- Intra-and inter-generational equity.

- Capacity-building and technology transfer.

- Environmentally sound techniques and methods of operation in order to prevent or limit damage to biological diversity.

- Sustainable use of marine biodiversity.

- Precautionary principle/approach.

- Risk-based approach.

- Polluter-pays principle.

- Special interests, circumstances and needs of developing countries such as SIDS and LDCs.

- Avoidance of disproportionate burden.

- Adaptive management.

- Ability to address cumulative impacts.

- Traceability.

- Flexibility.

- Conservation of biodiversity as a common concern of human-kind.

25. The impacts of climate change could be a consideration for decisions made and actions taken under the instrument and decisions should not exacerbate or hasten the adverse impacts of climate change, especially upon SIDS.

26. No action or activity taken on the basis of the instrument would be construed or considered to be prejudicial to the positions of States Parties to a land or maritime sovereignty dispute or to dispute concerning the delimitation of maritime areas.

B. *INTERNATIONAL COOPERATION*[1]

27. Cooperation, coordination, consultation and communication between and among States and international organizations, including regional and sectoral bodies, would be enhanced such as through exchange of information.

28. Guidance and recommendations could be provided to States,

---

[1] See also section Ⅳ.

including through existing global, sectoral or regional organizations involved in the conservation and sustainable use of marine biodiversity beyond national jurisdiction, in the form of goals, procedures, criteria, standards and guidelines.

29. Agreed general biodiversity protection guidelines or methodology to take into account the impact on fish stocks of emerging issues such as the adverse impacts of climate change, ocean acidification or pollution could be provided.

30. States would have a duty to cooperate directly or through appropriate subregional, regional or global mechanisms, taking into account the specific characteristics of the subregion or region (see UNF-SA article 8.1).

31. States Parties, in implementing the instrument, would be required to work together to actively engage competent international organisations and arrangements to take actions within their competence to contribute to the achievement of the objectives of the instrument. Where a State Party would consider that action is desirable in relation to a question falling within the competence of relevant international organisations or arrangements, it would be required to draw that question to the attention of the organisations or arrangement competent for that question. The Parties who are members of the organisation or arrangement in question would be required to cooperate within that organisation or arrangement in order to achieve an appropriate response.

32. Any State proposing that action be taken by an intergovernmental organization having competence with respect to marine biodiversity, where such action could have a significant effect on conserva-

tion and management measures already established by a competent sectoral or regional organisation or arrangement could be required to consult through that organisation or arrangement with its members or participants. To the extent practicable, such consultation would take place prior to the submission of the proposal to the intergovernmental organization.

33. An incentive for existing organizations to improve their performance could be created, including through the expansion of mandates to explicitly enable the adoption of measures to conserve marine biodiversity in areas beyond national jurisdiction, the addition of new principles to an existing organisation's governance framework, the development of processes within an existing organisation for implementing relevant parts of the instrument, or development of memorandums of understanding with other organisations to ensure better coordination.

34. If there is no body with a mandate for the conservation or sustainable use of marine biodiversity in a particular sector or geographic area of areas beyond national jurisdiction:

- *Option* 1: The establishment of a relevant body would be encouraged.

- *Option* 2: The establishment of a relevant body would not be encouraged as this would exceed the scope of the instrument.

35. In cases where there are a number of bodies but no effective coordination mechanism, effective coordination mechanism within a specific timeframe [article 8 (5) UNFSA refers] could be encouraged.

105

36. Formal or informal regional cooperative mechanisms could be strengthened or developed.

37. A possibility for participation of and/or cooperation with relevant international organizations, within their respective mandates, in practical arrangements under the instrument would be envisaged.

38. The duty of international organizations to coordinate and cooperate could be further operationalized as follows:

- Joint meetings where appropriate.

- Consultation on matters related to areas beyond national jurisdiction with a view to coordinate respective activities.

- Cooperation in the collection of data and information relating to areas beyond national jurisdiction.

- Sharing of information and data regarding activities and the impact of activities under their mandate with a scientific body under the instrument.

- Cooperation in the identification and implementation of the most effective conservation measures to protect areas in areas beyond national jurisdiction, within a specific timeframe.

- Cooperation in the management of MPAs in areas beyond national jurisdiction.

- Conduct of marine scientific research and joint assessments of the effectiveness of existing MPAs and their conservation measures.

- Regular reports to a conference of the parties on progress made.

- Participation in meetings of the respective governing bodies as observers.

C. *MARINE GENETIC RESOURCES, INCLUDING QUESTIONS ON THE SHARING OF BENEFITS*

1. *Scope*

39. Geographical scope:

* *Option* 1: The instrument would apply to marine genetic resources of the Area and the high seas.

* *Option* 2: The instrument would only apply to marine genetic resources in the Area.

40. Material scope:

* Fish and other biological resources used for research on their genetic properties.

○ A scientifically – informed threshold would be established, whereby if a particular (fish) species is extracted or harvested for the purpose of bioprospecting for marine genetic resources beyond a certain amount (depending on species and habitat variability), it would be considered a commodity. Such threshold could be elaborated by a scientific/technical body under the instrument.

* On *in situ/ex situ/in silico* resources:

○ *Option* 1: Applies to both *in situ* and *ex situ* marine genetic resources.

○ *Option* 2: Applies to *in situ* and *ex situ* marine genetic resources as well as *in silico* and digital sequence data.

○ *Option* 3: Applies to marine genetic resources collected *in situ*.

* On derivatives:

○ *Option* 1: Applies to derivatives.

107

○ *Option* 2: Does not apply to derivatives.

2. *Guiding principles and approaches*

41. On the common heritage of mankind and the freedom of the high seas:

• *Option* 1: The common heritage of mankind would underpin the new regime governing marine genetic resources of areas beyond national jurisdiction. This implies:

○ The need to carry out activities for the benefit of mankind as a whole, irrespective of their geographical location, and taking into particular consideration the interests and needs of the developing countries.

○ No claim or exercise of sovereignty or sovereign rights of the areas beyond national jurisdiction nor any appropriation would be recognized.

○ Benefits would be shared in a fair and equitable manner.

○ Activities regarding the exploration, exploitation of the resources in the said areas would be governed by the instrument.

• *Option* 2: The freedom of the high seas would be applied to marine genetic resources in areas beyond national jurisdiction.

• *Option* 3: Common concern of humankind.

• *Option* 4: No indication of the applicable legal regime.

42. The use of areas beyond national jurisdiction and their resources by all States would exclusively for peaceful purposes.

43. The jurisdiction and rights of coastal States over their continental shelf, including beyond 200 nautical miles where applicable, would need to be respected.

44. Principle of adjacency. Coastal States could be allowed a greater role in conserving, managing and regulating access to the resources of high seas pocket areas.

45. Marine scientific research activities do not constitute the legal basis for any claim to any part of the marine environment or its resources, as recognized in article 241 of UNCLOS.

46. Traditional knowledge.

47. Legal certainty, clarity and transparency.

48. Encouragement of research, innovation and commercial development.

49. Sustainable collection of genetic material.

50. Environmentally sound techniques and methods of operation.

51. Fair and non-arbitrary rules and procedures for benefit sharing.

52. Simple, expedient and cost-effective procedures and mechanisms.

3. *Access and benefit-sharing*

53. Existing access and benefit - sharing models that could be considered include:

- The provisions in UNCLOS relating to marine scientific research.
- Article 82 of UNCLOS.
- The CBD and Nagoya Protocol.
- The ITPGRFA.
- The Antarctic Treaty System.

54. A scientific/technical body under the instrument could elabo-

rate and recommend to the global body guidelines for the access and benefit-sharing regime.

3. 1 *Access to marine genetic resources of areas beyond national jurisdiction*

55. On whether to regulate access:

• *Option* 1: Free access to marine genetic resources could be provided, in line with UNCLOS provisions concerning marine scientific research in areas beyond national jurisdiction.

• *Option* 2: Regulated access could be provided:

○ *Option* 2. 1: For bioprospecting, not for marine scientific research purposes.

○ *Option* 2. 2: For marine genetic resources of the Area.

○ Terms and conditions for access could be established, taking into account the possibility of change of use, including capacity building, transfer of marine technology, a requirement to deposit samples, data and related information available in open source platforms such as databases, biorepositories and/or biobanks, and/or contribution to an access and benefit-sharing fund as conditions for access, drawn from the ISA model.

○ The principles contained in the Nagoya Protocol could be drawn from with respect to knowledge associated with genetic resources and prior consent involving indigenous and local communities.

○ Require States to take appropriate and effective legislative, administrative or policy measures to provide that genetic resources utilized within their jurisdiction have been accessed in accordance with established regulation.

110

56. *In situ* access to marine genetic resources of areas beyond national jurisdiction could be based, in particular, on the following:

• Respect for the regime for marine scientific research under UNCLOS.

• No hindrance to research and development.

• Respect for the rights and obligations of coastal States over the resources under their jurisdiction, as provided for in UNCLOS.

• Conservation and sustainable use of marine biological diversity of areas beyond national jurisdiction, in line with applicable UNCLOS provisions.

• Obligations for flag States to carry out the collection of marine genetic resources in a manner that does not harm the ecosystem and to use environmentally sound techniques and methods of operation.

• Impacts upon neighbouring zones, including those under national jurisdiction.

• Severity of impact upon adjacent States Parties.

• Consider cost of remediation and anti−pollution measures.

• Stricter environmental protection measures, if needed, in MPAs.

• Research programmes on marine research that could support the conservation and management of areas beyond national jurisdiction with the participation of developing countries inspired by article 16 of the ITPGRFA.

3. 2 *Sharing of benefits from the utilization of marine genetic resources*

3. 2. 1 *Objectives*

111

57. Contribute to conservation and sustainable use of marine biodiversity of areas beyond national jurisdiction.

58. Be beneficial to present and future generations.

59. Promote marine scientific research.

60. Enhance research and development.

61. Promote capacity building and technology transfer.

62. Build capacity to access marine genetic resources of areas beyond national jurisdiction.

3. 2. 2 *Principles guiding benefit-sharing*

63. Possible principles guiding benefit-sharing could include:

• Balance between the interests of participating States and other entities engaged in the access and use of marine genetic resources.

• Be fair and equitable on the basis of the common heritage of mankind.

• Be transparent.

• Be conducive to the conservation and sustainable use of marine biodiversity in areas beyond national jurisdiction.

• Not negatively impact States' rights to conduct marine scientific research consistent with the regime under UNCLOS.

• Be conducive to marine scientific research conducted in accordance with UNCLOS, as well as to the promotion of knowledge generation and innovation and not be detrimental to research and development.

• Give due consideration to SIDS and LDCs.

• Increase scientific knowledge on conservation of biodiversity.

• Benefits from humane use of derivatives and not promote or allow the use of derivatives towards destruction or impairment of human life or towards non-peaceful purposes.

3. 2. 3 *Benefits*

64. Types of benefits could include:

- *Option* 1: Both monetary and non-monetary.
- *Option* 2: Non-monetary benefits.

65. Possible monetary benefits could include those mentioned in the Annex to the Nagoya Protocol and in Part Ⅳ of the ITPGRFA.

66. Possible non-monetary benefits could include:

- Those mentioned in the Annex to the Nagoya Protocol.
- Those mentioned in Part Ⅳ of the ITPGRFA.
- Facilitation of marine scientific research.
- Collaboration in marine scientific research and in research and development programmes.
- Access to and dissemination of all forms of resources, samples, data and related knowledge, including through mechanisms for data sharing, such as a clearing-house, data banks, sample collections, and open access gene pools.
- Dissemination of research and development results relating to marine genetic resources.
- Collection and sharing of data and knowledge on the associated marine environment, biodiversity and ecosystems.
- The instrument could provide for a framework to specify, coordinate, promote and monitor the implementation, with respect to marine genetic resources from areas beyond national jurisdiction, of the provisions contained in Part ⅩⅢ (' Marine Scientific Research'),

113

such as promoting international cooperation in marine scientific research (article 242, UNCLOS), making knowledge resulting from marine scientific research available by publication and dissemination (article 244, para. 1, UNCLOS), and promoting data and information flow and the transfer of knowledge (article 244, para. 2, UNCLOS).

• Transfer of technology, including based on Part XIV of UN-CLOS.

• Capacity-building, including participation of scientists from developing countries in scientific research, access to scientific research vessels, educational opportunities and training programmes, activities to enhance, facilitate and stimulate the sharing of material, information and knowledge, strengthening capacities for technology transfer, institutional capacity-building, human resources and materials to strengthen capacities for the administration and implementation of access regulations, a global scholarship fund, and development of regional centres of excellence.

• Other socio-economic benefits (e. g. research directed to priority needs such as health and security).

67. The particular types of benefits that could be shared at particular points in the process could be considered.

3. 2. 4 *Benefit-sharing modalities*

68. A mechanism that takes into account existing benefit-sharing mechanisms such as the ISA could be considered.

69. A clearing-house mechanism could be established. ①

70. Possible benefit-sharing modalities could include:

• A fund could be established within the possible clearing-house mechanism:

○ The general purpose of the fund could be outlined in the instrument, whereas detailed modalities could be developed by the governing body or provided through a protocol.

○ Funded through royalties or milestones payments.

○ LDCs would be the primary beneficiaries of the fund.

○ Specific allocation for SIDS could be foreseen.

○ Benefits would be used for conservation and sustainable use of the ocean and marine biodiversity of areas beyond national jurisdiction.

○ Consideration could be given to whether special exemptions would be extended to developing and least developed countries where there would be obligations to pay into a benefit-sharing fund.

• A system similar to the annual partnership contribution under the World Health Organization (WHO) Pandemic Influenza Preparedness (PIP) Framework for the Sharing of Influenza Viruses and Access to Vaccines and Other Benefits could be established.

• International programmes on marine research could be established to support the conservation and management of the marine environment in areas beyond national jurisdiction. Inspired by the IT-PGRFA, Part Ⅴ (Supporting components), the research

---

① See sub-section 5 in this section.

programmes could include provisions on the participation of developing countries. The programmes could be linked to existing national research institutions and have research activities on genetic material from areas beyond national jurisdiction which may in turn generate monetary benefits for the participants. Another possibility would be to require developed Parties to include developing partners in their research and exploration activities on marine genetic resources.

- A hybrid mechanism by combining a project-based approach similar to the ITPGRFA Benefit-Sharing Fund could be established with inspiration drawn from the WHO PIP Framework.

3.3 *Intellectual property rights*

71. *Option* 1: Intellectual property rights, including disclosure of origin requirements in patent applications, would not be within the scope of the instrument, as this issue would need to be dealt with within the existing institutional frameworks competent in this subject-area [World Intellectual Property Organization (WIPO) and World Trade Organization (WTO)].

72. *Option* 2: A mandatory disclosure of origin of the marine genetic resource in patent applications or other intellectual property right would be established.

73. *Option* 3: The instrument would prohibit private appropriation and the exercise of intellectual property rights where this would limit access to marine genetic resources for further research and other aims. If intellectual property rights were claimed in respect of products developed from marine genetic resources, the approach taken in the ITPGRFA could be considered.

74. *Option* 4: The matter of intellectual property rights would have to be addressed in a manner that ensures consistency with the work being conducted under WIPO.

75. *Option* 5: A *sui generis* system would be developed.

4. *Monitoring of the utilization of marine genetic resources of areas beyond national jurisdiction*

76. Users could be required to register their activities.

77. A protocol, code of conduct or guidelines could be developed in order to ensure transparency in the use of marine genetic resources of areas beyond national jurisdiction.

78. A depository of information on marine genetic resources extraction could serve as a mechanism to trace the provenance of marine genetic resources obtained in areas beyond national jurisdiction.

79. A "passport" for marine genetic resources of areas beyond national jurisdiction could be introduced. This passport could be drawn from the Certificate of Compliance under the Nagoya Protocol, which would accompany genetic resources in order to demonstrate their origin at any stage of research, development, innovation, or commercialization.

80. A role for the ISA in monitoring of the utilization of marine genetic resources could be provided.

81. National authorities in charge of intellectual property rights could be established as a checkpoint to monitor the utilization of marine genetic resources and ensure benefit sharing.

5. *Clearing-house mechanism*①

82. The Nagoya Protocol Access and Benefit – Sharing Clearing House could be used as a model or expanded on to hold records and data related to marine biodiversity of areas beyond national jurisdiction.

83. The clearing-house mechanism could:

• Include global information services such as a website for the instrument, a network of experts and practitioners, mechanisms to exchange information and a network for clearing-house mechanisms at the regional, sub-regional and/or national levels.

• Include information on access to samples and sample collections, access to and transfer of technology and capacity building and funding opportunities, and data and knowledge sharing.

• Facilitate exchange of research results.

• Take into account the special circumstances of SIDS and LDCs.

84. The flag State could be required to report on accessed marine genetic resources from areas beyond national jurisdiction to the clearing house after the material has been deposited. A sample could also be provided to a public collection.

85. The deposit of sufficient information in a clearing – house could be required to enable use of marine genetic resources.

86. A platform for various organizations to cooperate and collaborate for better sharing of data/information could be created.

---

① See also sub-section 3. 2 in this section.

6. *Capacity building and transfer of marine technology*[①]

87. Proponents of marine genetic resources – related activities could be required to provide capacity building specifically to SIDS.

88. Elements of capacity building could include:

• Provision of education/training in science and technologies, policy and governance, including through joint research efforts supported through the establishment of a global scholarship fund, and enhanced through collaboration in research and development on marine genetic resources.

• A fund with specific allocation for SIDS.

D. *MEASURES SUCH AS AREA – BASED MANAGEMENT TOOLS, INCLUDING MARINE PROTECTED AREAS*

1. *Objectives of area – based management tools, including marine protected areas*[②]

89. Conservation and sustainable use of marine biological diversity of areas beyond national jurisdiction.

90. Protection and preservation of the marine environment.

91. Protection, maintenance and restoration of ocean health and resilience, including key ecosystem processes, habitats and species, and areas which are vulnerable to impact (s), including from climate change, such as unique, fragile/sensitive, rare or highly biodiverse habitats and features as well as areas essential for the sur-

---

① See also sub-section 3. 2 in this section.

② There is a need to determine whether it is practicable for the instrument to provide an exclusive list of objectives of ABMTs, including MPAs

vival, function, or recovery of particular stocks or rare or endangered marine species (such as breeding or spawning grounds), or for the support of large ecosystems.

92. Protection of representative marine ecosystems, biodiversity and habitats, including through a global, coherent and representative network of effectively managed ABMTs, including MPAs, in areas beyond national jurisdiction.

2. *Guiding principles and approaches*

93. Possible guiding principles and approaches could include:

• Consistency with UNCLOS, UNFSA and other relevant treaties.

• No undermining of existing relevant legal instruments and frameworks and relevant global, regional and sectoral bodies.

• Taking advantage of the work and expertise of existing relevant legal instruments and frameworks and relevant global, regional and sectoral bodies.

• Respect for the jurisdiction and rights of coastal States over the continental shelf, including beyond 200 nautical miles where applicable.

• Compatibility of measures taken for the EEZ and for areas beyond national jurisdiction.

• International cooperation and coordination.

• Necessity.

• Proportionality.

• Ecosystem approach.

• Precautionary principle/approach.

- Use of the best available science.
- Integrated approach.
- Preventive principle.
- Threats-based approach.
- Representativity.
- Adaptive management.
- Protection and preservation of the marine environment.
- Different levels of protection.
- Establishment and management of ABMTs, including MPAs, on an individual, case-by-case and temporary basis.
- Sustainable use.
- Balance of interests between activities for the protection and preservation of the marine environment and other lawful activities at sea.
- Multi-use approach.
- Equitable use.
- Transparency.
- Inclusivity.
- Public participation.
- Consultation.
- Accountability.
- Polluter-pays principle.
- Traceability.
- Liability.
- Stewardship of the marine environment for present and future generations.

121

- No disproportionate burden on coastal States.
- Traditional knowledge of local and indigenous communities.

3. *Process for the establishment of area-based management tools, including marine protected areas*

94. *Option* 1: "Global model" -A global overarching framework would be created to enable the identification, designation, management and enforcement of ABMTs, including MPAs, in areas beyond national jurisdiction.

95. *Option* 2: "Hybrid model" -General guidance and objectives would be developed at the global level to enhance cooperation and coordination and provide a level of oversight to the decision-making and implementation by regional and/or sectoral mechanisms.

96. *Option* 3: "Regional and/or sectoral model" -General policy guidance to promote cooperation and coordination would be provided at the global level, while recognising the full authority, without oversight from a global mechanism, of regional and sectoral organizations in decision-making.

97. The process would not be a " one - size - fits - all " approach. Rather, it would be on a "case-by-case" basis taking into account common guidelines established by the global instrument's institutions.

3. 1 *Identification of areas*

98. Areas requiring protection through ABMTs, including MPAs, could be identified on the basis of best available science.

99. Possible criteria for the identification of areas could comprise:

- General criteria and/or guidelines based on existing internationally recognized criteria such as those for CBD's Ecologically or Biologically Significant Marine Areas (EBSAs), IMO's PSSAs, vulnerable marine ecosystems (VMEs), and ISA's APEIs, including uniqueness, rarity, vulnerability, fragility, sensitivity, representativeness, dependency, naturalness, productivity and diversity.

- Criteria used under regional agreements.

- Contribution to a global representative network could be duly considered.

- Relevance of certain species or ecosystems for livelihoods and food security.

- Socioeconomic factors could also be at the core of the process of determining the size and location of ABMTs and MPAs.

- Criteria could be further developed, including on the basis of the best available scientific information and recommendations from a competent scientific advisory body.

- Any new criteria agreed to by States and a decision-making body under the instrument.

100. The work done by the IMO, CBD to describe EBSAs and RFMOs could be helpful in identifying priority biodiversity areas and vulnerable marine ecosystems.

3. 2 *Designation process*

3. 2. 1 *Proposal*

101. Possible proponents could include:

- States Parties to the instrument, individually or collectively, or in collaboration with relevant organizations.

• Those States and entities that would be entitled to become Parties.

• A scientific/technical committee under the instrument.

• Other intergovernmental organization （s） within their respective mandates.

• Non-governmental organizations.

102. Proponents would be encouraged to seek views and inputs from relevant stakeholders, including civil society, in the process of developing their proposals.

103. The proposal would need to take into account the best available science, the precautionary approach/principle, and follow an ecosystem approach.

104. Possible elements of a proposal could include:

• The delimitation of the area/spatial boundaries.

• A description of the objective （s）.

• A description of the characteristics and biodiversity values of the area and the sensitivity of the species/habitats concerned, including an evaluation of the current state of the marine ecosystem.

• A description of impacts, including cumulative impacts, identification of threats and possible activities with adverse impact.

• The conservation or management measures needed to reach the specified objective （s）, including human activities that would be banned, a management plan and socio - economic mitigation measures.

• A description, if relevant, of how the proposed site could contribute to ecologically representative MPA networks, including its

124

possible relationship to existing MPAs or other ABMTs.

- Information on neighbouring areas, including areas within national jurisdiction such as those covered by submissions under article 76 of UNCLOS.
- Information relating to possible interference with other legitimate uses of the sea and, when appropriate, related possible socio-economic costs.
- Information on international organizations and bodies whose action might be relevant in order to achieve the conservation objectives, including existing conservation measures by other international organizations.
- Whether prior consultations with those organizations have been held.
- Identification of any overlap between the proposed ABMT and an existing ABMT, including measures for coordination.
- Timing:

○ *Option* 1: A time period would be indicated for the conservation and management measures.

○ *Option* 2: The duration for the measures would not be indicated.

- Monitoring and review, including elements of a research and monitoring plan.
- Enforcement plan.

3. 2. 2 *Consultation on and assessment of the proposal*

A possible process for conducting consultations on and assessing the proposals could include the following:

105. A proposal could be circulated through the secretariat to:

• *Option* 1: States Parties and relevant global, regional or sectoral organizations and frameworks.

• *Option* 2: All States, including non - Parties, relevant global, regional or sectoral organizations, and civil society.

106. States, competent global, regional or sectoral organizations, and civil society representatives could be invited to submit comments on the proposal.

107. The agreement of States that are adjacent to areas beyond national jurisdiction where an ABMT could be created would be required when such an ABMT is under consideration.

108. There would be a time-bound period within which feedback and comments regarding the proposal could be submitted. The duration of the consultation process could be established in the instrument.

109. The contributions made during the consultation process could be made publicly available by the secretariat, which would also collect, compile and forward all comments back to the proponent.

110. The proponent would revise its proposal, as necessary, based on the comments received through the consultation process.

111. The proposal could be reviewed through a mechanism for scientific consideration and advice such as by a scientific/technical body, which would consider whether a similar MPA exists and how it could be complemented with protection under the instrument, provide advice on the proposal's compatibility with the instrument's scientific criteria, as well as make other recommendations, including on ecologically representative MPA networks and biogeographical classification schemes.

112. A scientific/technical body could make recommendations on the proposal.

3. 2. 3 *Decision-making*

In making a decision on the designation of ABMTs, including MPAs, the following could be considered:

113. The process of designation or establishment of ABMTs, including MPAs, would need to be consistent with the purposes and principles of the Charter of the United Nations, UNCLOS and other relevant legal instruments.

114. It would be necessary for the new instrument to provide for cooperation and coordination with existing regional and sectoral institutions. The relevant existing frameworks and non-contracting Parties would be allowed to join discussions of a global institution, such as a conference of the parties, as observers.

115. Measures taken by coastal States for the conservation and sustainable use of marine biodiversity within their national jurisdiction would need to be taken into account.

116. Adjacent States would be given particular consideration so that measures taken do not undermine their sustainable development.

117. The designation could include consideration of climate change and underwater noise impacts.

118. Depending on the conservation objectives of individual MPAs, as well as the vulnerability of their features/ecosystems and the pressures on them, different levels of protection could be necessary.

119. Time period for the ABMTs, including MPAs:

- *Option* 1: The designation would be for an indefinite period.
- *Option* 2: The designation would be time-bound.

- If, at a later stage, an MPA were to fall entirely within a maritime area under the sovereignty or jurisdiction of a coastal State, it would cease to be in force. The coastal State could decide to adopt similar measures under its national law. In case of partial overlap, the MPA's spatial boundaries would be amended accordingly.

120. Decisions on ABMTs would need to be based on scientific data with ABMTs being universal and binding in nature.

121. *Option* 1: A global institution would make decisions on:

- The spatial boundaries of the area to be designated as MPA.
- The establishment of such area.
- Appropriate conservation and management measures to be taken in the MPA.
- In taking the decisions, all efforts would need to be made to reach consensus. Majority voting could be envisaged.
- The ISA could be an essential component, as it has a mandate already recognized by UNCLOS.

122. *Option* 2: Information about a designated area and activities that could potentially harm or cause adverse impacts to that area would be referred to relevant bodies and frameworks with purview over such activities for consideration and possible management measures or other action by those bodies.

- When a designated area has multiple relevant bodies with purview over activities that could potentially harm or cause adverse impacts to that area, a process could be established by which those bodies could coordinate and cooperate, including during the consideration of and, as appropriate, implementation of possible management measures.

- If the existing framework decides to take different measures from those identified by the Conference of the Parties or none at all, the Conference of the Parties would ask the existing framework for consultations. The States Parties to the instrument and the relevant existing frameworks would cooperate as much as possible within the relevant frameworks to ensure these existing frameworks duly respect the decision of the Conference of the Parties and take appropriate conservation and management measures.

- Ways and means could be considered to make relevant measures binding upon all States Parties, including non-members of the relevant existing frameworks.

123. *Option* 3: Any issues related to the establishment and management of ABMTs, including MPAs, would be addressed within existing international mechanisms.

- An MPA proposal would be presented to the appropriate regional seas convention, which following public consultations, would consider the proposal in relation to the requirements of the instrument and relevant input from the public consultation.

- The MPA, including any measures that fall within its competence could be adopted by the regional seas convention which would announce its decision on its website. The decision made by the regional seas convention, in accordance with the requirements of the instrument, would be binding on all States Parties to the instrument.

129

- The regional seas convention, or one of its Parties, would forward the MPA decision specifically to other relevant bodies, such as IMO, RFMOs, other regional seas conventions, ISA, etc. The States Parties would be under an obligation to pursue the objectives of the instrument in all relevant mechanisms where they are participating.

- Other relevant bodies would consider whether the activity they are managing are relevant to the conservation goals and if measures within their competence would be required.

- The MPA would be on the agenda of the meeting of States Parties to the instrument following the decision, providing for accountability, transparency, review and stakeholder participation.

124. Where there is no competent body to recommend measures to address the impact of a specific activity in the proposed area:

- *Option* 1: The Parties would identify specific measures to meet the conservation objectives of the area.

- *Option* 2: Such measures would be developed and considered by Parties and those States and entities that would be entitled to become Parties.

- *Option* 3: The instrument would encourage the countries and organizations concerned to establish a new organization or framework and to participate in its activities.

125. Upon designation, a secretariat under the instrument would notify States Parties and non-Parties, relevant global, regional or sectoral organizations and frameworks, as well as all other stakeholders of the establishment of a new MPA, as well as its objectives, its management measures and the monitoring and review plan.

4. *Implementation*

126. MPA management measures would come into effect a certain number of days after adoption by the decision-making body, at which moment they would be binding on States Parties with respect to the processes and activities carried out under their jurisdiction or control, and on organizations.

127. It would be the responsibility of States Parties to the instrument to implement the management measures established with respect to activities and processes under their jurisdiction or control as flag States. The flag State and the port State concerned could cooperate in the implementation and enforcement of the MPAs.

128. Where the implementation of an MPA management measure would prevent a State Party from fulfilling an existing obligation under another relevant legal instrument or framework or by a relevant global, regional or sectoral body, the management measure would not enter into effect for the State Party concerned. Should this circumstance no longer apply, the MPA management measure would be considered to be binding on the State Party concerned.

129. Parties would commit to using their best efforts to ensure the adoption of necessary measures by the relevant organizations of which they are members.

130. Competent international organizations with mandate over the marine area concerned could be tasked with implementing and enforcing the ABMTS, including MPAs. They could also be invited to adopt specific measures necessary to achieve the conservation objectives of the new instrument. They could also be requested to de-

131

velop and implement biodiversity strategies and action plans as a tool to integrate biodiversity considerations into management and decision-making.

131. It would be important to ensure that area-based management measures would be made in a manner that takes account of action related to conservation and sustainable use of marine biodiversity taken by States in areas within national jurisdiction, and the interests of those coastal States adjacent to areas beyond national jurisdiction.

132. Nothing would prevent States Parties from adopting additional and stricter measures from those adopted by the competent international organizations with respect to their vessels or with regard to activities and processes under their control and jurisdiction.

133. While the management plan would not be applicable to non-Parties to the instrument, these would be notified of the designation and invited to consider implementing appropriate management measures for activities and processes under their jurisdiction having an impact on the conservation objectives of the MPA.

134. Any management measures and any enforcement of management measures would need to be consistent with UNCLOS, including but not limited to sovereign immunity (in line with article 236) and the obligations in article 237.

5. *Relationship to existing measures*

135. The relationship between provisions concerning ABMTs, including MPAs, in existing mechanisms and those to be made in the new instrument involves the relationship between existing international treaties and the instrument, which would need to be addressed in line

with the general principles concerning treaty application as provided for under the Vienna Convention on the Law of Treaties.

136. ABMTs, including MPAs, would not undermine existing MPAs and regulations that are implemented by relevant global, regional and sectoral bodies.

137. Nothing in the instrument would prejudice the ability of States to designate ABMTs, including MPAs, under other legal instruments and frameworks and relevant global, regional and sectoral bodies, or the obligations of States pursuant to such designations.

138. On a procedure of recognition of regional and sectoral ABMTs, including MPAs:

• *Option* 1: ABMTs, including MPAs, adopted by existing regional and sectoral mechanisms would go through a process of recognition by the global mechanism. Recognition would not derogate from the authority of a body to apply measures.

• *Option* 2: Sectoral ABMTs (e. g. RFMO VME closures, IMO PSSAs, or ISA APEIs) would not require a formal global recognition process, but the global body would be informed and the ABMTs included in the clearing – house mechanism and information – sharing mechanism.

6. *Capacity-building and transfer of marine technology*

Possible terms for capacity-building and transfer of marine technology could include:

139. Necessary support for developing countries, including obligations on developed States Parties to provide technical, scientific and funding support in the development of proposals, review of pro-

posals, development of management measures, and monitoring of ABMTs.

140. Provisions to avoid the transfer of disproportionate conservation burden on SIDS (e. g. modelled on article 7 of UNFSA).

7. *Monitoring and review*

Possible elements of monitoring and review could include the following:

141. The ABMTs, including MPAs, would be kept under regular review and be monitored on the basis of best available science against the objectives identified to assess their effectiveness.

142. A particular MPA's conservation and management plan and any specific measures applied to it could be adjusted to reflect the status of the area based on a review process.

143. On the role of a scientific/technical body:

• *Option* 1: A scientific/technical body under the instrument would oversee the monitoring and review of ABMTs, including MPAs.

• *Option* 2: This function could be delegated to regional bodies where possible and appropriate.

144. States Parties and competent global, regional or sectoral organizations could be required to report regularly on the implementation of the measures for activities under their purview. To this end, the instrument could provide for standardized reporting, with an associated timeframe for reporting, including through relevant regional and sectoral bodies.

145. Reports could be addressed to the secretariat that would make them available to a competent scientific advisory body for its

consideration and recommendations as appropriate, to all States Parties for consideration and decision if necessary, as well as to the general public, for information.

146. The monitoring and review process would take into account the scientific data and information provided by the scientific committee established under the new instrument, States, regional and sectoral bodies, as well as by relevant global and regional processes and frameworks (e. g. the Regular Process for Global Reporting and Assessment of the State of the Marine Environment, including Socioeconomic Aspects) and civil society. It could account for exogenous factors such as climate change.

147. The review process could result in the publication of a progress report and identify any shortcomings by Parties, non-Parties, and regional or global bodies, affecting the effectiveness of the measures.

148. The review could lead to the maintenance of the *status quo* established by the MPA, a modification of any of the parameters established, or the removal of the MPA. The conservation objectives and/or the management measures could be adjusted, if and when necessary, based on best available scientific information and recommendations from a competent scientific advisory body.

149. One way of mitigating displacement of destructive activities to other locations could be by giving a global governing body the mandate to monitor and assess such risks of displacement and introduce countermeasures.

150. There could be a mechanism under the instrument for a

135

global monitoring, control and surveillance (MCS) system to ensure that protected areas would be meeting their objectives and to identify violations by vessels as well as cases of regular non-compliance. This mechanism would facilitate information sharing and joint operations between existing MCS systems.

151. The instrument could establish a compliance mechanism. Following the outcome of the review process, Parties, stakeholders, including civil society, as well as the compliance committee itself, could submit a report of non-compliance. When a Party would be identified to have failed to discharge its obligations under the instrument or, in the case of non-Parties, under international law, to co-operate on the protection and preservation of the marine environment, by not taking measures or exercising effective control to ensure that its vessels or nationals do not engage in any activity that undermines the effectiveness of the instrument's conservation measures, the compliance committee would make recommendations on ways to rectify their acts or omissions. The non-complying Party and non-Party would be notified and offered a reasonable time to respond to the alleged non-compliance and rectify its actions or omissions. When necessary, the instrument would adopt measures to facilitate compliance (e. g. technical assistance and capacity building) based on recommendations from the compliance committee. If the Party or non-Party continues to undermine the effectiveness of the protected area, and/or if the ecosystem or any of its components under protection is under serious threat, the Parties to the instrument would adopt appropriate responsive measures. The responsive measures would be designed to

ensure that the conservation objectives of the area are met.

E. *ENVIRONMENTAL IMPACT ASSESSMENTS*

152. Internationally accepted standards, processes and protocols, including those reflected in the following instruments, could be referenced in developing the provisions on EIAs:

- UNCLOS.
- Convention on Environmental Impact Assessment in a Trans-boundary Context (Espoo Convention).
- Revised Voluntary Guidelines for the Consideration of Biodiversity in Environmental Impact Assessments and Strategic Environmental Assessments in Marine and Coastal Areas of the CBD (UNEP/CBD/COP/11/23).
- International Guidelines on Deep−Sea Fisheries in the High Seas of the Food and Agriculture Organization of the United Nations (FAO).
- ISA recommendations for the guidance of contractors for the assessment of the possible environmental impacts arising from exploration for marine minerals in the Area (ISBA/19/LTC/8).
- Protocol on Environmental Protection to the Antarctic Treaty.

1. *Obligation to conduct environmental impact assessments*

153. States would be required to conduct EIAs, as provided in article 206 of UNCLOS.

154. This obligation could be operationalized, including through the possible development of general guidance/guidelines.

- A scientific/technical body could elaborate and recommend guidance to a global decision−making body, including criteria and thresholds for activities in areas beyond national jurisdiction.

155. The obligation to conduct EIAs would rest with the State under whose jurisdiction or control the activity in question takes place, namely where the State exercises effective control over a particular activity or the State exercises jurisdiction in the form of licensing or funding a particular activity, and not simply activities conducted by a vessel flying a State's flag.

156. The EIA could be carried out by a third party, such as a research institution or a private company, under the direction and control of the State.

2. *Guiding principles and approaches*

157. EIAs would contribute to the conservation and sustainable use of marine biological diversity of areas beyond national jurisdiction.

158. Possible guiding principles and approaches could include:

- Precautionary principle / approach.
- Ecosystem approach.
- International cooperation.
- Integrated approach.
- Use of the best available science / science-based approach.
- Transparency.
- Inclusiveness.
- Consultation.
- Fairness.
- Effectiveness.
- Inter-and intra-generational equity.
- Responsibility to protect and preserve marine environment.

- Polluter-pays principle.

- Stewardship.

- No-net-loss principle.

3. *Activities for which an environmental impact assessment would be required*

159. The obligation to conduct EIAs would relate to planned activities under the jurisdiction or control of States.

160. Possible approaches to set out when an EIA would be required could include:

- *Option* 1: EIAs would be mandatory for all proposed activities in areas beyond national jurisdiction.

- *Option* 2: EIAs would be required under specified circumstances, including based on:

○ Possible threshold levels could be:

- *Option* 1: Based on UNCLOS article 206 ("reasonable grounds to believe that a proposed activity may cause significant and harmful changes to the environment").

- *Option* 2: More stringent requirements than UNCLOS to include "any harmful" changes.

- *Option* 3: "Minor or transitory impact" as a preliminary threshold requiring initial assessment to determine whether significant impacts would be likely and formal EIA and reporting would be required as a result.

- *Option* 4: More than a "minor or transitory effect".

- For areas designated for application of ABMTs under the instrument, or otherwise designated for their significance/vulnerability

139

at the international level (e. g. , EBSAs, VMEs, PSSAs, MPAs),
a specific threshold could be determined.

- Threshold could be included in an annex.

○ Approaches to listing activities could include:

- *Option* 1: Develop an indicative list of activities that would require EIAs (cf. Espoo Convention, Appendix Ⅲ).

○ A list would be non-exhaustive and not legally-binding.

○ A list would be placed in an annex.

- *Option* 2: Develop a list of activities exempt from EIAs.

- A list could be developed by the institution with responsibility
for guiding the conduct of EIAs.

- A list could be reviewed or updated by a conference of the
parties to reflect new and emerging uses and scientific and
technological developments.

○ Where an activity in areas beyond national jurisdiction is already covered by existing obligations and agreements, possible approaches could include:

- *Option* 1: It would be unnecessary to conduct another EIA for
these activities under the instrument.

- *Option* 2: No such activity would be seen as exempt by definition.

- Activities in areas designated for application of ABMTs under
the instrument, or otherwise designated for their significance/vulnera-
bility at the international level (e. g. , EBSAs, VMEs, PSSAs,
MPAs), would require an EIA.

161. EIAs would not be related to global processes (for example,

ocean acidification, global warming), which depend on many factors, and regulation of which is currently carried out within the relevant competent international structures.

162. Cumulative impacts could be considered:

* *Option* 1: Possible cumulative effects would be taken into account, including those resulting from climate change, ocean acidification, and deoxygenation that may increase the significance of the effect of proposed projects.

* *Option* 2: Cumulative impacts would be assessed "as far as practicable".

4. *Environmental impact assessment process*

163. Possible general procedural steps for the conduct of EIAs could include:

* Screening

○ *Option* 1: Decision on the need for an EIA would be made by the State party under whose jurisdiction the proponent operates.

○ *Option* 2: A global body under the instrument would articulate when or how activities in areas beyond national jurisdiction trigger the need for an EIA.

* Scoping

* Assessment and evaluation of impacts

○ An assessment of the potential impacts of the proposed activities in every dimension would need to be taken into account.

○ The related evaluation and analysis of the risks and potential impacts or effects of the proposed activities to marine environment would be done on the basis of recognized scientific methods.

- Reporting of the EIA

○ Consistent with articles 205-206 of UNCLOS, the reports of the results of the assessments would be published and communicated.

- Review / Monitoring

○ Consistent with article 204 of UNCLOS, States would be required to keep under surveillance the effects of any activities being undertaken following the positive outcome of any EIA.

164. Possible approaches to public participation/involvement could be as follows:

- *Option* 1: Consultation would exist at each stage of the EIA process, beginning with the scoping phase.

- *Option* 2: The type and frequency of public notification and consultation would reflect an activity's level of risk and its anticipated impacts.

- *Option* 3: Stakeholders would have an opportunity to provide inputs before decisions are made.

165. Stakeholders for the conduct of public consultation could be:

- Adjacent coastal States that may be affected by the impacts of the activity.

- States Parties to the instrument.

- Regional or sectoral bodies existing in the area where the activity would be conducted.

- Relevant intergovernmental and non-governmental organizations.

- Relevant experts from the scientific community and/or a scientific and technical committee created by the instrument.

- Affected industries.

166. The relevant State (s) could circulate a draft assessment that includes public comment and input and the information required by the instrument to all relevant stakeholders.

167. The proponent of a proposed activity in areas beyond national jurisdiction would be required to notify the State under whose jurisdiction the proponent falls.

168. A scientific/technical body could:

- Oversee the EIA process.

- Review proposals and provide recommendations to a global decision-making body regarding submissions on EIAs, including an assessment of cumulative impacts of human activities in areas beyond national jurisdiction and proposed provisions of an environmental management plan, including monitoring, review and compliance provisions.

- Carry out periodic and ex-post evaluations.

- A pool of experts capable of conducting EIAs for activities in areas beyond national jurisdiction could be created under a scientific/technical body and could be commissioned to conduct and evaluate EIAs by States with capacity constraints.

169. An SEA/EIA administrative oversight committee could:

- Establish guidelines for EIAs.

- Ensure that the EIA and SEA processes would be properly conducted by the appropriate entities and provide advice to a global body on EIAs and SEAs.

170. Where relevant international organizations have competence as regards EIAs in areas beyond national jurisdiction within specific

143

sectors and/or areas, States could be required to conduct EIAs either directly or through relevant global, regional or sectoral bodies.

171. The draft assessment, along with any subsequent comments and recommendations, would be made publicly available on a website or equivalent.

172. Any completed EIA would be included in a report to a global body. States Parties, relevant bodies, non-governmental organizations, etc.would be given the opportunity to evaluate and scrutinize the assessments, considerations and decisions.

173. A decision on whether to proceed with proposed activity could be made by:

• *Option* 1: The State Party under whose jurisdiction or control the activity takes place.

• *Option* 2: An international body created under the instrument, such as a conference of the parties, with the advice of a technical and scientific committee under such a body, and with an appeals process.

174. It would be necessary to ensure that the outcome of the EIA is duly taken into account in decisions on the authorization of activities and on any accompanying mitigation or compensation (redress) measures.

175. The proposed activity would be permitted only where the assessment concludes that the activity would not have significant adverse impacts, or could be managed to avoid such impacts. Each decision to permit an activity would include an environmental management plan.

176. When an activity would not be authorized, an appeals process could be provided for.

177. Neither the EIA itself nor the State's decision based on the

EIA would be subject to review by any outside entity or process.

178. On the question of who would bear the costs of EIAs:

● *Option* 1: The instrument could address who would bear the costs for an EIA.

○ The costs of the EIAs could be borne by or contributed to by the operator.

○ The costs of conducting the EIA could be borne by the proponent of an activity.

○ In the case of activities carried out by developing countries, consideration could be given to the need for financing and/or other means of cooperation (capacity-building and technology transfer).

● *Option* 2: The decision on costs could be left to the national competence of States Parties.

5. *Content of environmental impact assessment reports*

179. The content of EIA reports could include:

● A description of the proposed activity and its purpose.

● A description of the environment likely to be affected, including any dependent or associated ecosystems, ecosystem services provided, impacted sensitive or vulnerable areas and vulnerability to climate stressors.

● A description of the ecosystem services provided by the area.

● A description of the potential environmental impact of the proposed activity, including impacts on ecosystem services.

● Cumulative, direct, indirect, short-term and long-term, positive and negative effects.

● A description, where appropriate, of reasonable alternatives to the proposed activity, including the non-action alternative.

● A description of mitigation measures to keep adverse environmental impacts to a minimum.

● Baseline information.

● An indication of predictive methods and underlying assumptions, as well as the relevant environmental data used, and an identification of gaps in knowledge and uncertainties encountered in compiling the required information.

● Follow-up actions to verify the accuracy of the EIA and the effectiveness of mitigation measures, including where appropriate, an outline for monitoring and management programmes and any plans for post-activity analysis.

● A rehabilitation plan, if necessary.

● Enforcement and compliance provisions.

● A non-technical summary.

180. A generic EIA template could be developed.

6. *Environmental impact assessments for transboundary impacts*

181. Article 206 of UNCLOS could serve as a basis for assessing transboundary impacts.

182. Transboundary impacts would not require separate assessment processes.

183. With regard to activities in areas within national jurisdiction having impacts beyond national jurisdiction, the following would need to be considered

● *Option* 1: All human activities with the potential for significant adverse impacts in areas beyond national jurisdiction would need to be assessed, regardless of where they actually take place.

● *Option* 2: The instrument would not cover EIAs for those activities within national jurisdiction.

○ States Parties would be required to develop national legislation to cover activities within areas under their national jurisdiction and to publish reports.

184. Activities in areas beyond national jurisdiction that have a potential impact upon the areas or resources within national jurisdiction would be subject to a transboundary EIA.

185. When an activity in areas beyond national jurisdiction would have an impact on an adjacent coastal State:

● *Option* 1: That coastal State would be given due attention in the conduct of a project planning and EIA, including through consultation with the coastal State.

● *Option* 2: That coastal State would be notified and allowed to be intimately involved in the EIA process, particularly the evaluation, and the activity would not be allowed to proceed without the specific approval of the affected coastal State.

● Communities would also be notified and consulted.

● Civil society could also be notified and consulted.

7. *Strategic environmental assessments*①

186. On whether or not to include a provision on SEAs:

● *Option* 1: The object of EIAs would be the planned "activities" under the jurisdiction or control of States, excluding SEAs.

① See also sub-sections 4 and 12.

- *Option* 2: The conduct of SEAs would be provided for.

○ Clear, transparent and effective requirements and procedures for SEAs would be established.

○ The parameters for EIAs in the instrument would equally apply to SEAs.

○ Cooperation between States at the regional level would be facilitated, either *ad hoc* or in the context of existing regional or global institutions, for conducting SEAs in areas beyond national jurisdiction.

- SEAs would be developed at the regional level, prior to commencing activities requiring EIAs.

- Regional and global organizations would be encouraged to prepare SEAs where they have mandates.

- Mechanisms would be established in the instrument to engage global sectoral organizations, such as IMO and ISA, as well as regional conventions in regional SEA processes.

○ SEAs would be collectively funded.

8. *Compatibility of environmental impact assessment measures*[1]

187. Compatibility with coastal State measures could be built into EIAs conducted in relation to areas beyond national jurisdiction adjacent to areas within coastal State's jurisdiction.

9. *Relationship to existing environmental impact assessment measures under relevant legal instruments and frameworks and relevant global, regional and sectoral bodies*

Possible approaches to EIA measures under relevant instruments

---

① See also section 6.

and frameworks and relevant global, regional and sectoral bodies could include the following:

188. Existing relevant legal instruments and frameworks, in particular UNCLOS, as well as relevant global, regional and sectoral bodies should not be undermined.

189. Existing processes and guidance developed to assess the impacts of human activities on biodiversity features applicable in areas beyond national jurisdiction, including those under regional and sectoral regimes would be respected.

190. There would be no duplications between EIAs conducted under the instrument and EIAs conducted under the relevant existing bodies.

191. Existing activities managed under regional and sectoral organizations could be allowed to continue where these organizations are mandated to consider the environmental impacts in the regulation of their respective activities. The instrument could play a useful role in assisting to coordinate these efforts.

192. Cooperation and information sharing between the different conventional regimes that envisage an EIA would be facilitated.

193. The global body under the instrument could assure transparency, accountability and stakeholder scrutiny of assessments and decisions made.

10. *Clearing−house mechanism*

194. Information gathered in connection with an EIA process in areas beyond national jurisdiction would be made publicly available.

195. A central repository of publicly available data and

149

information on EIAs, SEAs, and baseline data on areas beyond national jurisdiction, such as a clearing-house mechanism could be established, including to:

- Publish draft EIAs.
- Provide for stakeholders to comment on the draft EIAs within a set deadline.
- Communicate the results of EIAs.
- Contribute to capacity-building, in particular for developing countries, including by facilitating access to a globally shared body of best practices.

196. A clearing-house would, as far as possible, be cost-effective, take advantage of information technologies, including through a dedicated website managed by DOALOS.

11. *Capacity-building and transfer of marine technology*

197. The special needs of developing countries, in particular SIDS, LDCs and land-locked developing countries would need to be taken into account, including necessary technical, knowledge and financial assistance, as well as development of infrastructure, institutional capacity and transfer of marine technology, amongst others.

198. Adequate capacity-building and transfer of marine technology would need to be ensured, including through collaboration, e. g. voluntary peer review mechanisms or "twinning" amongst States Parties.

199. The special circumstances of SIDS would need to be adequately addressed.

- The mechanism would need to ensure that the required

capacity would be available for SIDS, so that the contents of the EIAs and associated implications are fully understood and comprehended.

• Financial and technical support to develop and review EIAs and to encourage more equity in activities in areas beyond national jurisdiction could be included.

200. Developing countries could be provided with an opportunity to submit joint EIAs where appropriate.

12. *Monitoring and review*

Possible approaches to monitoring and review could include the following:

201. Consistent with article 204 of UNCLOS, States would monitor and keep under surveillance the effects of any activities undertaken following the EIA, as well as compliance with any conditions (such as prevention, mitigation or compensation measures) related to their authorization.

202. Monitoring and review could be performed as follows:

• *Option* 1: A monitoring and review mechanism would ensure compliance.

○ On an annual basis, States Parties would be required to prepare and submit to a review committee a report detailing their implementation of the EIA-related provisions of the instrument. States could also report on any failures to implement the EIA-related provisions by other parties. The reports would be made publicly available without delay.

○ With the assistance of the secretariat and the scientific body, the committee would prepare an annual synthesis document evaluating

151

States' compliance with their EIA – related obligations, identifying any specific instances of non–compliance and publish such report.

○ Affected coastal States and relevant regional/sectoral bodies would be consulted by the monitoring and compliance committee in the conduct of monitoring and evaluation activities.

• *Option* 2: Monitoring and review would be performed by the State or the proponent of an activity with regular reporting to the State concerned.

203. After termination of the activity, there would be a follow–up evaluation to ensure environmental protection was upheld, which could take the form of natural capital accounting and be compared against the baseline established during the screening phase.

204. A contingency fund could be established to mitigate possible harmful effects on the environment caused directly by the activity. In line with the polluter–pays principle, proponents of the activity would deposit an agreed sum of money, which would be returned to the proponent upon satisfactory completion of an ex–post EIA and clearance from the scientific committee of the global body.

205. Reports about subsequent measures and monitoring results would be made publicly available.

F. *CAPACITY–BUILDING AND TRANSFER OF MARINE TECHNOLOGY*[①]

206. The instrument could define general obligations in promoting

---

[①] Capacity – building and transfer of marine technology could be mainstreamed throughout the instrument or be included in a standalone section.

cooperation to develop capacity and transfer of marine technology, including in the following manner:

- *Option* 1: Capacity−building and transfer of marine technology are cross cutting issues and affect all the other issues in the package deal, and thus would be mainstreamed into the other sections of the instrument.

- *Option* 2: A dedicated section would be included which would focus on the various elements with links to the other sections.

207. Capacity−building and technology transfer provisions would be coherent with and enhance the implementation of or operationalize existing provisions on capacity−building and transfer of marine technology of UNCLOS and other international agreements.

208. The instrument could include provisions with reference to the following:

- UNCLOS, Part XIV.
- CBD, article 18. 1.
- Criteria and Guidelines on the transfer of marine technology of the Intergovernmental Oceanographic Commission (IOC) of UNESCO.
- Small Island Developing States Accelerated Modalities of Action (SAMOA) Pathway (paras. 102 and 111).
- Stockholm Convention on Persistent Organic Pollutants, articles 11 and 12, paras. 1 and 2.
- Minamata Convention on Mercury, article 14. 1.
- Istanbul Programme of Action for the Least Developed Countries for the Decade 2011−2020 (IPoA).

1. *Objectives of capacity−building and transfer of marine technolo-*

gy

209. The instrument could provide for general and specific objectives relating to capacity－building and transfer of marine technology for the conservation and sustainable use of marine biological diversity of areas beyond national jurisdiction. These could include：

• Improving the capacity of developing countries in the conservation and sustainable use of marine biological diversity of areas beyond national jurisdiction.

• Increasing, disseminating and sharing knowledge and expertise on the conservation and sustainable use of marine biodiversity in areas beyond national jurisdiction, and empowering all States to fully take part in the achievement of the instrument's objectives.

• Coordinating efforts relating to the conservation and sustainable use of marine resources.

• Enhancing and developing the capacity of developing countries to implement the instrument.

2. *Principles and approaches guiding capacity － building and transfer of marine technology*

210. Possible guiding principles and approaches could include：

• Duty to cooperate and collaborate under UNCLOS.

• Duty to promote the development of marine scientific and technological capacity of States under UNCLOS.

• Duty to provide scientific and technical assistance to developing countries under UNCLOS.

• Provision of data and information on the basis of best available science.

- Needs-driven and meaningful.

- Long-term support.

- Pertinence.

- Effectiveness.

- Equality.

- Mutual benefit.

- Transparency.

- Integrated approaches.

- Duty to provide preferential treatment for developing countries under UNCLOS.

- Special regard to the requirements of developing States (similar to obligations under Part VII of UNFSA) . This duty would include consideration of the countries with special interests and needs:

○ SIDS.

○ LDCs.

○ Land-locked developing countries.

○ Geographically disadvantaged countries.

○ Coastal African States.

○ Coastal communities vulnerable to the impacts of climate change.

○ Specific challenges of developing middle-income States.

- Promote the role and participation of women.

- Consider the principles included in the IPoA.

- Involve relevant stakeholders, including private actors and organizations.

155

- Part XIV of UNCLOS would provide a basis for capacity-building and transfer of marine technology, and the IOC Criteria and Guidelines on Transfer of Marine Technology would provide a basic framework for capacity-building and transfer of marine technology.

- Enhance the implementation of and build upon lessons learned from existing instruments and mechanisms, including UNCLOS, ISA, IOC-UNESCO, United Nations Framework Convention on Climate Change (UNFCCC), CBD, IPoA, without undermining or duplicating them.

- Optimize the use of available financial, human and technical resources.

211. As regards the relationship between intellectual property rights and capacity-building and transfer of marine technology:

- *Option* 1. Ensure the protection of intellectual property rights.

- *Option* 2. Give due regard to intellectual property rights.

- *Option* 3. Strike a balance between the protection of intellectual property rights and the promotion and dissemination of technology, including by referring to intellectual property in the organizations that are competent in such issues, in particular WIPO or WTO Trade Related Aspects of Intellectual Property Rights.

3. *Scope of capacity-building and technology transfer*

212. Both capacity-building and transfer of marine technology could address:

- Access, collection, analysis and use of data, samples, publications and information.

- Implementation of UNCLOS obligations to promote the development of marine scientific research capacity in developing States and to promote the transfer of marine science and technology.

- Benefits from developments in marine science related activities.

- Capacity-building in respect of access and benefit sharing. ①

- Development, implementation and monitoring of ABMTs, including MPAs. ②

- Conduct and evaluation of EIAs, and participation in SEAs. ③

- Implementation of the Sustainable Development Goals, in particular Sustainable Development Goal 14.

3. 1 *Capacity-building*

213. With regard to the types of capacity-building activities for inclusion in the instrument, the following could be considered:

- *Option* 1. A list would not be included given that it might be too prescriptive and could hamper the ability to adapt to future developments. A general requirement would be included in the instrument leaving details to be possibly determined at a later stage by an *ad hoc* working group.

- *Option* 2. An indicative, non-exhaustive and flexible list of activities would be incorporated. It could include the following:

○ Development of human resource and institutional capacity, through initiatives at the regional, sub-regional and national levels

---

① See also section C.

② See also section D.

③ See also section E.

across sectors and organizations to implement the instrument.

○ Individual capacity-building through short-term, medium-term and long-term training and scholarships, exchange of experts.

○ Scientific, educational, technical assistance, including in natural and social sciences, both basic and applied, including oceanography, chemistry, marine biology, marine geospatial analysis, ocean economics, international relations, public administration, policy and law, training in science and technologies, including through the establishment of a global scholarship fund.

○ Assistance in the development, implementation and enforcement of national legislative, administrative or policy measures, including associated regulatory, scientific and technical requirements on a national or regional level.

○ Establishment or strengthening of the capacity of relevant organizations/ institutions.

○ Access to and acquisition of necessary knowledge and materials, information, and data in order to inform decision making of the developing countries.

○ Awareness-raising and knowledge sharing, including on marine scientific research.

○ Development of joint research cooperation programmes, technology in marine science, necessary infrastructure, acquisition of necessary equipment to sustain and further develop R&D capabilities in country, including data management.

○ Collaboration and international cooperation in scientific research projects and programmes.

158

○ Establishment or strengthening of the capacity of relevant organizations/ institutions.

3. 2 *Technology transfer*

214. Any definition of transfer of marine technology would need to be broad enough to take account of future developments in science.

215. The IOC Criteria and Guidelines on the Transfer of Marine Technology provide an important reference point, and details could be included on what is considered technology for the purpose of technology transfer, with the possibility for revision to meet the requirements of the instrument.

216. Technology transfer could include the following:

• Access to technology that is appropriate, reliable, affordable, modern and environmentally sound.

• Hard technology as well as other associated aspects such as computers, autonomous underwater vehicles and remotely operated underwater vehicles.

• Specialized equipment, such as acoustic and sampling devices, multi – beam echo sounding, acoustic underwater positioning systems.

• Observation facilities and equipment, *in situ* and laboratory observations such as analysis and experimentation, molecular tools for high – resolution observation of microbes to larger invertebrates that would allow sequencing of DNA at sea and back on shore.

• IT infrastructure that would allow advanced data analysis and storage of data, including high – resolution, large – scale and long – term data collection.

159

● Data and specialised knowledge inclusive of, but not limited to, equipment, manuals, sampling methodology, criteria, reference materials, guidelines, protocols, samples, processes, software, methodologies and infrastructure.

● Institution building at the regional, sub-regional and national levels, including for the management of data.

● Training and technical advice and assistance necessary to assemble, maintain and operate a viable system and the legal right to use these items for that purpose on a non-exclusive basis.

● Innovative financial mechanisms for marine technologies.

4. *Modalities for capacity-building and technology transfer*

217. Capacity-building and technology transfer could be provided as follows:

● Through clear, simple, targeted procedures and modalities that operate as expeditiously as possible, by working directly or through appropriate global, regional organizations and bodies.

● On a case-by-case basis, country specific and needs driven, to provide tailored solutions for States requiring it.

218. Capacity-building and transfer of marine technology would need to be responsive to national and regional needs, priorities and requests, with flexibility to adapt to changing needs and priorities. Needs could be evaluated through:

● Periodic assessments of the needs identified by developing States carried out at national and regional levels, involving all relevant States and stakeholders.

● A holistic evaluation of existing capacities, including institutional and human resource capacities.

- Data from the Sustainable Development Goals indicators.

219. Transfer of marine technology could be provided as follows:

- *Option* 1: On fair and reasonable terms and conditions as well as through favourable terms and conditions.

- *Option* 2: On a voluntary basis, on mutually agreed terms and conditions that respects intellectual property rights and fosters science, innovation, research, and development.

- *Option* 3: On a voluntary basis, on favourable terms, including on concessional and preferential terms, as mutually agreed.

220. Cooperation at all levels would be important and could be facilitated through:

- North–South, South–South and triangular cooperation and partnerships with relevant stakeholders, including intergovernmental organizations, non-governmental organizations, academia, the business sector/private sector and philanthropic organizations.

- Collaboration between Regional Seas Programmes and RFMOs.

- Development of joint scientific research projects in cooperation with institutions in developing countries and the establishment of national and regional scientific centres of excellence, including as data repositories.

- Sharing knowledge for, and raising awareness on the importance of effective conservation and sustainable use of marine biodiversity in areas beyond national jurisdiction.

- Joint venture arrangements and advisory and consultative services which enable human resources development, education, technical assistance/cooperation, development and transfer of technology.

221. Development of human resources as well as technical and research capabilities related to the objectives and material scope of the instrument could be effected through the following:

● Creation of training opportunities at national, regional and global levels, including exchange postings and workshops.

● Establishment of mentoring and partnerships.

● Development of regional centres for skill development.

● Establishment of a global scholarship programme to foster science, policy and governance research on the conservation and sustainable use of marine biological diversity of areas beyond national jurisdiction in a similar manner to the United Nations – Nippon Foundation of Japan Fellowship Programme.

● Development of a strong global professional alumni network as a pool of human resources, networking, mutual learning, and a foundation of international cooperation.

222. Best practices and lessons learned from existing mechanisms would need to be utilized wherever relevant and applicable, including:

● The mechanism under the ISA.

● CBD, article 16.

● The Agreement on Port State Measures to Prevent, Deter and Eliminate Illegal, Unreported and Unregulated Fishing.

5. *Clearing-house mechanism*

223. A global system, linking clearing-house mechanisms networks at the global, regional and national levels and providing a central "one-stop shop" access to information could be established.

224. The interoperability and linkages between existing clearing-house mechanisms could also be improved.

225. A clearing-house mechanism could perform the following functions:

• Provide a platform or repository, including as a centralized information access point for the dissemination of, sharing and coordination of knowledge, including traditional knowledge, data and information, technological activities, including access to evaluations and publications.

• Help to ensure quick/one-stop access to information on capacity-building and technologies in relation to the objectives and scope of the instrument.

• Promote and facilitate access to corresponding expertise and know-how, including through virtual classes.

• Provide information on existing opportunities and projects, activities and programmes occurring in areas beyond national jurisdiction and a method for matching needs and opportunities for capacity-building and transfer of marine technology.

• Identify best practices and recognize gaps to better support the implementation of the instrument.

• Develop initiatives at national, regional and global levels.

• Promote international coordination and collaboration.

163

- Facilitate open access to samples and knowledge.

226. A Secretariat or other institutions could be in charge of administering the clearing-house mechanism.

227. A clearing-house mechanism could build on and not duplicate existing instruments, mechanisms and frameworks, including the following:

- UNCLOS, Part XIV.

- The IOC-UNESCO Criteria and Guidelines on Transfer of Marine Technology, the International Oceanographic Data and Information Exchange and the Ocean Biogeographic Information System. The relationship between the instrument and IOC could be clarified, including whether to enhance the role of IOC through additional financial support or resources to provide and develop a structure for fostering coordination and collaboration.

- The work of ISA.

- The Nagoya Protocol.

- The UNFCCC and the Paris Agreement.

- The ITPGRFA Global Information System.

- The Convention on Access to Information, Public Participation in Decision-Making and Access to Justice in Environmental Matters (Aarhus Convention).

6. *Funding*

228. A funding mechanism (s) to ensure adequate, predictable and sustainable funding for capacity-building and transfer of relevant marine technology, as well as to promote the establishment of genuine partnerships between the private sector and private and public actors

in developing countries, could be established as follows:

- *Option* 1: A voluntary trust fund would be established.

- *Option* 2: An existing funding mechanism would be utilized, for example the Global Environment Facility.

- *Option* 3: A special fund and other distinct funding mechanisms such as a rehabilitation or liability fund, as well as a contingency fund would be established.

- *Option* 4: A combination of voluntary and mandatory mechanisms.

229. Funding would be provided through:

- Voluntary and mandatory proceeds. Existing funding mechanisms such as the Nagoya Protocol, and ISA capacity – building funding arrangements could be models to draw from.

- Contributions resulting from the access to and utilization of marine genetic resources; premiums paid during the approval process for EIAs; penalties incurred for non-compliance for EIAs; and a percentage of the amount paid for the transfer of technology.

- Contributions from sponsoring States or private entities proposing to explore and exploit marine biological diversity resources of areas beyond national jurisdiction, with rates of contribution depending on considerations such as the size of area involved, type of activities, and risks associated with the proposed activities.

230. The funding mechanism could be integrated with the climate change mechanism, and similar funding mechanisms, for instance taking into account carbon footprints.

231. Contributions to the fund would be open to Member States,

165

other entities as well as non-governmental organizations, foundations, research centres, individuals, etc.

232. New ocean sustainability finance tools could be considered, such as the Coalition for Private Investment in Conservation.

233. The fund could be used to fund capacity-building and transfer of marine technology related activities and programmes, including:

- Finance the participations of developing countries in major meetings under the instrument.

- Assist developing countries in meeting their commitments under the instrument.

- Support scholarships and fellowships, programmes, training, and other opportunities for nationals of developing countries to learn about activities related to marine biological diversity of areas beyond national jurisdiction and participate fully in the operationalization of an instrument.

- Support regional scientific and technological centres with pooled global resources to enhance technology transfer efforts.

- Support the development a clearing-house for capacity-building and transfer of marine technology.

234. Any funding mechanism would need to have minimal conditionality for access and use of funds.

235. The resources for capacity-building and technology transfer would need to be promptly received by the target State.

236. Priority access to a fund and preferential treatment could be given to SIDS and LDCs.

166

237. The fund could have dedicated earmarking for vulnerable States.

7. *Monitoring, review and follow-up*

238. A monitoring, review and follow-up process could:

• Enable the review on a periodic basis of the capacity constraints faced by developing countries, in particular SIDS, so that the recipient countries and regions' needs could be adequately met, on a stable and long-term basis.

• Measure the success of capacity – building and technology transfer efforts, utilizing quantitative and qualitative data, carried out in a joint collaborative effort undertaken at the national, regional and global level.

239. A monitoring, review and follow – up process could be carried out through the following:

• An advisory (scientific and/or technical) or decision-making body under the instrument.

• A review conference and/or meeting of the States Parties could be convened on a regular basis to assess the needs and to fill in the gaps, supported by a Secretariat and/or a compliance committee.

• States Parties could be made aware of the progress made in capacity-building under the instrument.

• A review process that would be inclusive of all stakeholders which contribute to capacity-building and transfer of marine technology.

240. Reporting requirements could be established that would be regular, transparent, comprehensive and streamlined for SIDS and facilitate periodic and systematic reviews, including of needs and priorities.

IV. INSTITUTIONAL ARRANGEMENTS

241. The instrument could provide for institutional arrangements as follows:

- *Option* 1: "Global model" – Scientific advice, decision – making, review and monitoring of implementation would be done at the global level.

- *Option* 2: "*Hybrid model*" –General guidance, criteria and standards would be set at the global level while regional and sectoral organizations would be relied upon for scientific advice and implementation and compliance, with a level of oversight as regards decision – making and implementation at the global level.

- *Option* 3: "Regional or sectoral approach" –A global mechanism would aim at facilitating coordination and cooperation while leaving regional and sectoral bodies with the full authority to decide on measures and ensure follow–up and review of implementation.

242. The possibility of using mechanisms already in place could be examined.

A. *DECISION–MAKING BODY/ FORUM*

243. The institutional arrangement for the instrument would

provide for an overarching framework at the global level that would meet regularly. It could be organized as follows:

- *Option* 1. A new international organization, with a meeting or conference of the parties would be established. The conference of the parties would be convened every year and a Review Conference[1] every five years.

- *Option* 2: The mandate of the ISA would be expanded to oversee the implementation of the instrument.

244. The global body could be composed of an assembly and a council, with limited membership whose members would be elected by the assembly.

245. Possible functions of a global body could include:

- Establish objectives, procedures, criteria, standards and guidelines, based on best available scientific information, including traditional knowledge.

- Oversee/review the implementation of the instrument.

- Adopt decisions related to the implementation of the instrument.[2]

- Establish subsidiary bodies and provide guidance to these bodies, as considered necessary.

- Facilitate and promote cooperation and coordination among different stakeholders, States and competent organizations, including by establishing processes for cooperation and coordination with existing

---

[1] See also section XI.

[2] See e. g. sections III. C, III. D, III. E and III. F.

bodies.

- Administer a global information repository.

- Consider and adopt amendments to the instrument.

- Promote harmonization of appropriate policies and measures for the conservation and sustainable use of marine biodiversity of areas beyond national jurisdiction.

- Ensure compliance with the instrument.

- Adopt programmes of work and budgets relating to the work of the instrument.

- Assess the effectiveness of the instrument in securing the conservation and sustainable use of marine biodiversity in areas beyond national jurisdiction, and if necessary, propose means of strengthening the implementation of the instrument in order to better address any continuing problems in the conservation and sustainable use of marine biodiversity.

- Review information received from States Parties and relevant sectoral and regional mechanisms on actions taken with regard to the implementation of the instrument.

- Adopt decisions on how to undertake implementation in the absence of a competent regional body or where such a body fails to take action.

- Consider any other issues as decided by the Parties.

246. The regulations and measures put in place by the global body "shall be no less effective than international rules, standards and recommended practices and procedures", in line with UNCLOS article 208 (3).

247. Participation in meetings could be open to non–Parties, relevant inter–governmental organizations, non–governmental organizations and other stakeholders, in an observer capacity.

248. The global body could welcome input from existing regional and sectoral organizations, civil society, and other stakeholders as appropriate.

249. The meetings of the global body could be held at a venue at which most delegations, and in particular SIDS, maintain a permanent presence, in order to take fully into account their particular capacity constraints.

250. Some decision–making as well as implementation could be conducted at the regional level, so as to adequately reflect regional and sub–regional specificities.

251. States Parties would be encouraged to, where possible, co-operate through regional instruments with an objective to implement measures adopted under the new instrument.

252. A regional/sub – regional forum could be established and could meet at regular intervals, prior to the meeting of the global body, to:

- Take decisions on measures to implement based on global criteria, standards and measures.
- Organize broad and inclusive consultations with relevant stakeholders on relevant projects.
- Report to the global body.
- Make recommendations or submissions for improving the implementation of the instrument to the global body.

253. The regional/sub-regional forum could be composed of two chambers (adjacent coastal States; and all parties to the instrument) and its meetings could be open to representatives of existing regional organizations, existing sectoral organizations, international organizations, and other stakeholders.

254. Where a subregional or regional organization or arrangement exists and has the competence to establish conservation and sustainable use measures, States Parties to the instrument could be required to become a member of such organization in order to effectively cooperate in such organization or arrangement, and actively participate in its work.

255. A State Party which is not a member of a subregional or regional organization or arrangement or is not a participant in a subregional or regional organization or arrangement could nevertheless cooperate, in accordance with relevant international agreements and international law, in the conservation and management of the relevant fisheries resources by giving effect to any conservation and management measures adopted by such organization or arrangement.

256. Representatives from relevant organizations, both governmental and non-governmental, concerned with biological diversity beyond areas of national jurisdiction could be afforded the opportunity to take part in meetings of subregional and regional organizations and arrangements as observers or otherwise, as appropriate, in accordance with the procedures of the organization or arrangement concerned. Such representatives may be given timely access to the records and reports of such meetings, subject to the procedural rules on

access to them.

B. *SUBSIDIARY BODY/BODIES*

257. A scientific and/or technical body could be established as a subsidiary body, including as follows:

• *Option* 1. A scientific and/or technical body would be established. It could utilize scientific committees under existing frameworks. It could be organized in chambers or sub-commissions similar to the Commission on the Limits of the Continental Shelf.

• *Option* 2. One scientific committee covering all sea areas would be established.

• *Option* 3. Multiple scientific committees with each one covering a sea area would be established.

258. The possible composition of these bodies could include:

• Multidisciplinary subject-matter experts nominated by governments, including from States Parties on issues covered by the instrument.

• Representatives and international experts specializing in various elements of the instrument, for example from the FAO and the IMO.

• Experts in or relevant traditional knowledge holders.

259. A scientific and/or technical subsidiary body could perform the possible following functions:

• Decision-making, this could be by consensus in principle.

• Making recommendations to the global body in relation to marine genetic resources, including questions on the sharing of benefits,① ABMTs, including MPAs,② EIAs,③ and capacity building and transfer of marine technology. ④

• Identifying new and emerging issues relating to the conservation and sustainable use of marine biodiversity of areas beyond national jurisdiction.

• Providing advice on scientific programmes and international cooperation in research and development related to conservation and sustainable use of marine biodiversity of areas beyond national jurisdiction.

• Responding to scientific, technical, technological and methodological questions that the decision-making body and its subsidiary bodies might submit.

• Providing regular assessments of the state of scientific knowledge of marine biodiversity of areas beyond national jurisdiction. Due consideration could be given, among others, to the question of what kind of input could be received from the Regular Process, as well as other relevant processes (such as the EBSA process).

• Carrying out additional functions such as those of a financial, budgetary and legal nature, as deemed necessary.

260. Additional subsidiary bodies could be established under the instrument as follows:

---

① See also section Ⅲ. C.
② See also section Ⅲ. D.
③ See also section Ⅲ. E.
④ See also section Ⅲ. F.

- An SEA/EIA administrative oversight committee. [1]

- A compliance committee to review general issues of compliance and implementation of the instrument. [2]

- A finance and administration committee. [3]

- A committee on capacity-building and transfer of marine technology.

- A mechanism/entity with a mandate to oversee access and benefit-haring of marine genetic resources. [4]

261. Taking into account the special case of SIDS, each of the subsidiary bodies could allocate dedicated seats to SIDS.

262. At the regional level, regional arrangements could be established to facilitate implementation of the instrument, including regional experts panels or committees, such as regional area-based management committees, regional capacity-building and transfer of marine technology committees, regional enforcement committees, and regional finance and administration committees.

C. SECRETARIAT

263. The functions of the secretariat could be modelled on article 319 (2) of UNCLOS and article 15 of General Assembly resolution 49/28, as well as the general functions of secretariats under other instruments, such as the CBD and UNFCCC.

264. Procedures established by the secretariat could take into ac-

---

[1] See also section III. E.
[2] See also section VII.
[3] See also section VI.
[4] See also section III. C.

count the special circumstances of SIDS.

265. Whether a permanent secretariat would be required or whether secretariat services could be provided by an existing international body, such as secretariat the Division for Ocean Affairs and the Law of the Sea, Office of Legal Affairs (DOALOS/OLA) services would need to be provided in a cost-effective manner. DOALOS/OLA could serve as secretariat for the instrument with the necessary allocation of human, technical and financial resources.

266. The secretariat and the depositary may not necessarily be the same.

V. EXCHANGE OF INFORMATION/ CLEARING-HOUSE MECHANISM

267. Exchange of information and data would be promoted between States as well as relevant regional, sectoral and global organizations (similar to article 17 of CBD).

268. Clear information would be provided and principles established to allow meeting papers, meeting reports, decisions, annual reports and results of any performance monitoring of the organization to be made available in a timely manner to Parties, civil society and outside institutions.

269. The instrument could include an article on transparency, taking a similar approach as in article 18 of the Convention on the Conservation and Management of High Seas Resources of the South Pacific Ocean. Alternatively, requirements to promote/ensure transparency could be integrated throughout the instrument.

270. States and subregional or regional biological diversity management organizations and arrangements could give due publicity to conservation and sustainable management measures and ensure that laws, regulations and other legal rules governing their implementation would be effectively disseminated.

271. A clearing-house mechanism could be established and perform the following functions:

- Promoting and facilitating the sharing of information, knowledge and data.

- Promoting and facilitating technical and scientific cooperation.

- Maintaining a network of experts and practitioners among Parties and partners.

- Gathering information on:

○ Marine genetic resources and the data deposited by entities
that obtain permits to access said resources.

○ Benefit-sharing in its monetary and non-monetary forms,
possibly including payments and financial resources.

○ Scientific data regarding ABMTs and EIAs, as well as follow
-up reports and related decisions taken by competent bodies.

○ Capacity-building and transfer of marine technology opportunities and offers.

- Linking to regional and national clearing-house mechanisms.

272. Regional clearing-house mechanisms would be part of the
global clearing-house mechanism.

273. The clearing-house mechanism could be managed by the
secretariat.

274. An evolutionary approach could be employed where the in-
formation-sharing functions would be carried out by the secretariat
until such a time when the extraction of marine genetic resources be-
comes a reality, at which point a specific body would be established.

VI. FINANCIAL RESOURCES AND MECHANISM

A. *FUNDING MECHANISM*

Possible approaches to funding could include the following:

275. Funding to support the implementation of the instrument could be provided through:

- Mandatory sources (contributions from States Parties and royalties and milestone payments from exploitation of marine genetic resources).

- Voluntary contributions from States Parties, States non-Parties, international financial institutions, donor agencies, intergovernmental organizations, non – governmental organizations; and natural and juridical persons.

276. A global trust fund could be established to perform the following functions:

- Fund the participation of developing States Parties in the instrument's processes.

- Assist developing countries in meeting their commitments under the instrument, including through conduct of EIAs.

- Fund capacity-building activities.

- Fund technology transfer-related activities and programmes, including training.

- Support the conservation and sustainable use programmes by holders of traditional knowledge in local communities, including in areas within national jurisdiction, so as to support coherence in ocean management.

- Support public consultations at the national and regional levels.

277. An endowment fund, managed by the secretariat, could promote and encourage the conduct of collaborative marine scientific research in areas beyond national jurisdiction including research activities related to marine genetic resources in these areas, by supporting the participation of qualified scientists and technical personnel from developing countries in marine scientific research programmes and activities and by providing opportunities to these scientists to participate in relevant initiatives.

278. Possible approaches to ensure that the special case of SIDS would be taken into account could include:

- Providing for a SIDS specific allocation in the fund (s).

- Providing for a special SIDS procedure with a pre-application process, which could then trigger a support mechanism to prepare the required application.

- Making use of existing funding mechanisms.

279. It would be necessary to ensure that procedures for access to funding and reporting would not be burdensome.

280. A finance and administration committee could carry out the

following possible functions:

- Draft financial rules, regulations and procedures.

- Assess contributions of Parties.

- Draft rules, regulations and procedures on the equitable sharing of financial and other economic benefits derived from marine genetic resources and the decisions to be made thereon.

- Facilitate resource mobilisation for implementation of the instrument and provide assistance to Parties, especially developing countries, and among those, particularly LDCs and SIDS.

- Review and plan budget.

- Monitor the funds established in the instrument.

- Report to the global body.

B. *REHABILITATION / CONTINGENCY FUND*

Possible approaches to address rehabilitation and contingencies could include the following:

281. A mechanism to deal with loss, damage and contingencies could be developed drawing on experience from the Warsaw Mechanism for Loss and Damage established under the UNFCCC, and other similar regimes.

- Such mechanism would have a residual nature, i. e. to enter into action only when the primary entities liable or responsible could not completely deal with the damage or rehabilitation need.

- Clear criteria would be established for the funding, which could come from an enterprise's up – front payment, sponsoring States' deposit of a bond, voluntary contributions, mandatory contributions, or a mix of all of these.

181

282. In line with the polluters – pay principle, a rehabilitation fund could be established. Private entities wishing to engage in the exploration and exploitation of marine biodiversity of areas beyond national jurisdiction would be required to contribute to the fund, in accordance with a scale tied to the degree of potential environmental harm stemming from activities related to marine biodiversity of areas beyond national jurisdiction of those entities. The fund would be used to finance the rehabilitation of marine biodiversity of areas beyond national jurisdiction, including their natural environments, in the event of pollution or other damaging impacts on marine biodiversity of areas beyond national jurisdiction and/or the areas beyond national jurisdiction in which they reside.

283. A contingency fund could be established to finance environmental disasters, such as pollution and other catastrophic disasters caused by human activities.

VII. IMPLEMENTATION

284. States and all those engaged in management of biological diversity could be required, for areas under the instrument, to adopt harmonised measures for the long–term conservation and sustainable use of biological diversity.

285. States, individually or collectively, including through sectoral and regional organizations where these exist, would be responsible for implementation, as well as ensuring compliance and enforcement of their flag vessels, nationals, and entities under their jurisdiction, with respect to the instrument.

286. States would be required to enact legislation and regulations and/or adopt measures necessary to ensure compliance with the standards, measures and procedures set up in the instrument.

287. Possible approaches to monitoring and reviewing compliance could include:

- Developing a global MCS system for areas beyond national jurisdiction to facilitate information sharing and joint operations between existing MCS systems.

- Possible institutional arrangements could include:

○ *Option* 1. The global body could be in charge of monitoring and reviewing compliance with the instrument, and enforce its provisions whenever they may be breached.

- Compliance could be the focus of a specifically mandated body established by the conference/meeting of parties. It could be established as follows:

  ○ The committee could be composed of a facilitative branch that would provide advice and assistance to Parties in order to promote compliance and an enforcement branch that would determine consequences for Parties not meeting their commitments.

  ○ The non-compliant Parties and non-cooperating non-Parties would be notified and offered a reasonable time to respond to the alleged non-compliance and rectify their actions or omissions.

  ○ Non-compliance complaints by non-State actors could be received by this body for further analysis and brought to the attention of the global body for appropriate follow up.

  ○ *Option* 2. Regional and sectoral bodies could be in charge of monitoring and reviewing compliance with the instrument.

288. Any mechanism for implementation and enforcement would take account of regional bodies.

289. A regular reporting and review process could be set up whereby:

- Parties and relevant regional or global bodies would report back regularly on the implementation of the instrument and conservation and management measures. These reports would be publicly available.

- Input from the scientific committee, all relevant regional or global bodies, and stakeholders, including civil society, as well as information gathered through the global MCS system would be provided.

• The review process would publish a progress report and identify any shortcomings by Parties, non-Parties, and regional or global bodies, affecting the effectiveness of the measures adopted by the instrument.

VIII. SETTLEMENT OF DISPUTES

Possible approaches to dispute settlement could include the following:

290. There could be a dispute prevention mechanism to pre-empt any dispute from arising. Such issues could be examined either by a specific committee or by selected experts.

291. States would be required to resolve their disputes relating to the interpretation and application of the instrument by peaceful means.

292. The parties could consider submitting the case to a third-party procedure based on explicit mutual agreement.

293. Consideration could be given to the means contained in Article 33 of the Charter of the United Nations. The 1982 Manila Declaration on the Peaceful Settlement of Disputes could also serve as a model.

294. With regard to the provisions of UNCLOS relating to the peaceful settlement of disputes:

- *Option* 1: The provisions reflect a good starting point for consideration of dispute resolution under the instrument.

- *Option* 2: It would not be appropriate to directly apply these provisions.

186

295. The provisions of UNFSA on dispute settlement could be used as a model.

296. Jurisdiction could be given to the International Tribunal for the Law of the Sea over contentious disputes, as well as advisory powers. A special chamber to deal with issues related to marine biodiversity of areas beyond national jurisdiction could be created.

297. A new body using the International Tribunal for the Law of the Sea as model could be established.

298. Regional dispute settlement mechanisms could also be considered.

299. Consideration could be given to the inclusion of qualified opt-out mechanisms modelled on the South Pacific Regional Fisheries Management Organization's Convention mechanism, where a measure could go forward and a State opting-out could have recourse to arbitration over the matter.

IX. NON-PARTIES

300. Non-Parties could be encouraged to become parties to the instrument.

301. States that would not be Parties to the instrument would not be discharged from their general obligations under UNCLOS and customary international law, including the obligation to protect and preserve the marine environment as well as the obligation to cooperate in good faith, and to ensure that their activities do not undermine the effectiveness of the instrument's conservation measures.

302. The relevant rules of the Vienna Convention on the Law of Treaties would apply.

X. RESPONSIBILITY AND LIABILITY

303. Possible approaches to responsibility and liability include:

• *Option* 1: A provision similar to article 35 of UNFSA would be included.

• *Option* 2: A provision would be included which reflects and builds upon the responsibility of States under international law to not cause damage to areas beyond national jurisdiction or to other States, by "ensur [ing] that activities within their jurisdiction or control do not cause damage to the environment of other States or of areas beyond the limits of national jurisdiction".

• *Option* 3: No provision on responsibility would be included in the instrument as the articles on Responsibility of States for Internationally Wrongful Acts elaborated by the International Law Commission and attached to General Assembly resolution 56/83 represent an authoritative body of international law in this field.

304. Guidance could be drawn from the polluter−pays principle, the International Law Commission Articles on Transboundary Harm from Hazardous Activities, as well as conventional regimes addressing liability.

305. In determining the scope of the liability, guidance could be drawn from Section 21 of the draft regulations on exploitation of mineral resources in the Area of the ISA.

306. The 2011 Advisory Opinion of the International Tribunal on the Law of the Sea on responsibilities and obligations of States sponsoring persons and entities with respect to activities in the Area could inform the instrument.

XI. REVIEW

307. A possible mechanism for regular review of the effectiveness and implementation of the instrument could be established, similar to the review mechanism set out in article 36 of UNFSA. Reviews could be carried out, based on agreed criteria, within a set period of time after entry into force of the instrument, for example after five years, and regularly thereafter.

308. A short period of time between the reviews would need to be ensured.

309. Reviews could address the following aspects:

• Performance of the institutional body set up under the instrument in carrying out its designated functions.

• Decisions taken under the instrument against the objectives, principles and standards set out in the instrument.

• Performance of Parties to the instrument in terms of implementation of the instrument.

• Performance of regional and sectoral bodies with a role in the conservation and sustainable use of marine biodiversity in areas beyond national jurisdiction in fulfilling their functions under the instrument.

XII. FINAL CLAUSES

310. The instrument would contain standard final clauses, such as those contained in articles 37 to 50 of UNFSA and 309 to 319 of UNCLOS, including provisions relating to settlement of disputes, signature, ratification and accession, entry into force, reservations and exceptions, declarations and statements, amendment, denunciation, participation by international organizations, depository, and authentic texts.

311. Consideration would need to be given to the number of ratifications required for entry into force, ensuring a prompt entry into force.

312. With regard to participation:

• Universal participation would be sought. The instrument would be open for signature, ratification and accession by all States and other entities on the same basis as provided for in UNFSA (articles 37-39).

• Similarly to article 305 in connection with Annex IX of UNCLOS, the instrument would also be open for signature by international organizations allowing for the participation of the European Union.

• Consideration could be given to the necessity and possibility of provisional application of the instrument.

192